DARWINISM: SCIENCE OR PHILOSOPHY?

Copyright © 1994 by
Foundation for Thought and Ethics
Richardson, Texas

Printed in the United States of America

ISBN 0-9642104-0-1

Library of Congress Catalog Card Number: 94-94523

DARWINISM: SCIENCE OR PHILOSOPHY?

Proceedings of a symposium entitled: "DARWINISM: SCIENTIFIC INFERENCE OR PHILOSOPHICAL PREFERENCE?"

Held on the Southern Methodist University campus in Dallas, Texas, March 26-28, 1992.

Sponsored by the Foundation for Thought and Ethics, Dallas Christian Leadership, and the C.S. Lewis Fellowship.

EDITORS

Jon Buell
and
Virginia Hearn

Foundation for Thought and Ethics • Richardson, Texas

Contents

SYMPOSIUM PARTICIPANTS

Phillip E. Johnson - J.D., University of Chicago, Jefferson E. Peyser Professor of Law, Boalt Hall, University of California, Berkeley. Author of *Darwin on Trial* (Second Edition, Downers Grove, IL: InterVarsity, 1993) and *Evolution as Dogma: The Establishment of Naturalism* (Dallas: Haughton Publishing, 1991).

Peter Van Inwagen - Ph.D. in Philosophy, Professor of Philosophy, Syracuse University. Author of *Material Beings* (New York: Cornell University Press, 1992).

Michael E. Ruse - Ph.D. in Zoology, Professor of Zoology and Philosophy of Science, University of Guelph. Author of *Darwinism Defended* (Reading, Mass: Addison-Wesley, 1982), chief expert witness in McLean v. Arkansas (the "Arkansas Creation Trial").

Arthur M. Shapiro - Ph.D. in Zoology, Professor of Zoology and Entomology, University of California, Davis. Acting Book Review Editor of the journal *Evolution*, author of over 200 published works.

Michael J. Behe - Ph.D. in Chemistry, Associate Professor of Chemistry, Division of Biochemical Sciences, Lehigh University. Author of numerous professional articles.

David L. Wilcox - Ph.D. in Population Genetics, Professor of Biology, Eastern College, St. David's, PA. Author of *The Creation of Species by Means of Supernatural Selection,* (submitted to Baker Book House) and numerous professional articles.

William A. Dembski - Ph.D. in Mathematics, M.A., Ph.D. candidate in Philosophy. Post-doctoral Fellow in Philosophy of Science, Nothwestern University, 1992-93. Director, Center for Interdisciplinary Studies in Princeton.

Leslie K. Johnson - Ph.D. in Zoology, Lecturer of Ecology and Evolutionary Biology, Princeton University. Author of numerous professional articles.

K. John Morrow, Jr. - Ph.D. in Genetics, Professor of Biochemistry and Molecular Biology, Texas Tech University Health Sciences Center. Author of numerous professional articles.

Frederick Grinnell - Ph.D. in Biochemistry, Professor of Cell Biology and Neuroscience, University of Texas Southwestern Medical Center in Dallas. Author of *The Scientific Attitude* (New York: Guilford Press, 1992), and over 100 professional articles.

Stephen C. Meyer - Ph.D. in Philosophy of Science, Assistant Professor of Philosophy, Whitworth College. Author of several articles on science and public policy.

Introduction

THERE ARE SOME THINGS that we expect all rational, educated persons to believe, regardless of their philosophical or religious standpoint. That the earth is roughly spherical, and revolves around the sun, is one of those things. We do not call persons who still believe in a geocentric universe "dissenters"; we call them cranks.

The literature of evolutionary biology is full of statements to the effect that something called the "fact of evolution" is as certain as that the earth goes around the sun. And so it is—if by "evolution" we mean only that selective breeding produces new varieties, or that island species have differentiated from mainland species, or that living things have certain common features (e.g., the genetic code) that suggest *some* process of development from *some* common source.

Evolution is a word of many meanings, some of which are controversial and some of which are not. One meaning of evolution that is highly controversial is *Darwinian* evolution (or the neo-Darwinian synthesis), when it is offered as a general description of how life progressed from very simple beginnings to its present complexity and diversity. As I describe in my opening address, "Darwin's Rules of Reasoning," the philosophically important doctrine of evolution is what I call the "blind watchmaker thesis," after the famous book by Richard Dawkins. This thesis argues not simply that life evolved to its present state of complexity and diversity, but that it evolved by the purposeless material mechanisms of random genetic change and natural selection. The implication is that humanity is a cosmic accident produced by a mindless cosmos.

That naturalistic explanation of how life came to its present state of complexity and diversity is a major prop for naturalism in philosophy and for agnosticism in religion. The question discussed at this symposium is whether the theory of evolution also receives essential support from that same philosophy, in which case there is a certain circularity of reasoning. The call to the meeting put the point at issue in this way: "*Darwinism and neo-Darwinism as generally held in our society carry with them an a priori commitment to metaphysical naturalism, which is essential to make a convincing case on their behalf.*"

1

I put the same question differently in my exchange with Michael Ruse: "Is there any reason that a person who believes in a real, personal God should believe Darwinist claims that biological creation occurred through a fully naturalistic evolutionary process?"

I do not think the issue was ever really confronted on this question, because the Darwinists tended to shift the focus to a different question. What the anti-Darwinists called metaphysical naturalism the Darwinists called "science," and they insisted that for science to cease being naturalistic would be for it to cease being science. To put the matter in the simplest possible terms, the Darwinist response to the question presented was not "No, that is wrong, because the case for Darwinism can be made without assuming a naturalist perspective." Instead, they answered "So what? All that you are really saying is that Darwinism is science."

To put this response in the words of the participants, Michael Ruse defined "scientific methodology" as including "a commitment to the idea of the world being law-bound—that is, subject to unbroken regularity—and to the belief that there are no powers, seen or unseen, that interfere with or otherwise make inexplicable the normal workings of material objects." Replying to Ruse for the anti-Darwinists, Stephen Meyer observed that "Either brute matter has the capability to arrange itself into higher levels of complexity or it does not, and if it does not, then either some external agency has assisted the arrangement of matter or matter has always possessed its present arrangement." Michael Ruse argued that science is limited by its nature to the first of these possibilities. Stephen Meyer argued that a rational historical biology must not arbitrarily limit the possibilities for consideration at the outset. The argument was repeated throughout the sessions, and no agreement was reached.

If that were all there were to say, the symposium might have seemed an exercise in futility. But that is not all there is to say. Academic conferences do not usually end with all philosophical disputes resolved in agreement, and they are not judged as failures if the participants go away with their basic commitments unchanged. What a successful conference does is to enable the participants and spectators to understand the issues better, and especially to give the participants a better understanding of each other as people. Judged by that standard, the symposium at Southern Methodist University was a tremendous success, perhaps even an historic event.

To understand why, it is necessary only to reflect on the state of the debate over evolution and creation. Historically, it has taken the form of a debate over the authority of Genesis (a subject this symposium did not touch upon), and it has been carried on in the bitterest terms. Scientists, including those who participated on both sides of the issue at SMU, have come to regard creation/evolution debates as circuses at best, and occasions for deception or manipulation at worst. The organizers of this symposium—Jon Buell, Stephen Sternberg, and Thomas Woodward— had to work hard at overcoming those suspicions for the symposium to occur at all. Understandably, things began on a wary note.

By the end of the symposium all that was in the past. What had been demonstrated is that it is possible for basic metaphysical differences to be debated with good will and humor; disagreement over even this issue does not have to lead to war. By the end of the second day the participants were socializing and conversing with gusto. I recall that every single participant at one time or another expressed the hope that other similar conferences would be held in the future. The healthy atmosphere penetrated the host institution to the extent that faculty members who kept away from the conference out of wariness have invited me back a year later for a faculty colloquium.

The SMU symposium was never meant to resolve the great debate over evolution and creation. It was meant to set the conditions so such a debate could begin. It was meant to set an agenda for debates in the future, with papers of the highest professional quality dealing with the scientific and philosophical issues in dispute. Given the doubts we all had at the beginning, the fact that the symposium accomplished so much seems almost a miracle—if I may use that term without offending the metaphysical naturalists who helped make the miracle possible.

That sets the general picture, but even scientific materialists are apt to remind us that "God is in the details." Those philosophical and scientific details are in the papers that follow, and in the vigorous responses to each paper from the opposing side. Individually, they illuminate specific areas of dispute. Taken together, they prepare the reader for a great debate that will occupy the attention of many minds during the coming decade.

Phillip E. Johnson
Berkeley, California
January 1993

I
Keynote Presentations

First Keynote Address
Darwinism's Rules of Reasoning
Phillip E. Johnson

Second Keynote Address
Darwinism: Philosophical Preference, Scientific Inference, and Good Research Strategy
Michael Ruse

Response to Michael Ruse
Laws, Causes, and Facts
Stephen C. Meyer

1
Darwinism's Rules of Reasoning
Phillip E. Johnson

MY STARTING POINT is a book review that Theodosius Dobzhansky published in 1975, critiquing Pierre Grassé's *The Evolution of Life*.[1] Grassé, an eminent French zoologist, believed in something that he called "evolution." So did Dobzhansky, but when Dobzhansky used that term he meant neo-Darwinism, evolution propelled by random mutation and guided by natural selection. Grassé used the same term to refer to something very different, a poorly understood process of transformation in which one general category (like reptiles) gave rise to another (like mammals), guided by mysterious "internal factors" that seemed to compel many individual lines of descent to converge at a new form of life. Grassé denied emphatically that mutation and selection have the power to create new complex organs or body plans, explaining that the intra-species variation that results from DNA copying errors is mere fluctuation, which never leads to any important innovation. Dobzhansky's famous work with fruitflies was a case in point. According to Grassé,

> The genic differences noted between separate populations of the same species that are so often presented as evidence of ongoing evolution are, above all, a case of the adjustment of a population to its habitat and of the effects of genetic drift. The fruitfly (*drosophila melanogaster*), the favorite pet insect of the geneticists, whose geographical, biotropical, urban, and rural genotypes are now known inside out, seems not to have changed since the remotest times.[2]

Grassé insisted that the defining quality of life is the intelligence encoded in its biochemical systems, an intelligence that cannot be understood solely in terms of its material embodiment. The minerals that form a great cathedral do not differ essentially from the same materials in the rocks and quarries of the world; the difference is human intelligence, which adapted them for a given purpose. Similarly,

> Any living being possesses an enormous amount of "intelligence," very much more than is necessary to build the most magnificent of cathedrals. Today, this "intelligence" is

called information, but it is still the same thing. It is not programmed as in a computer, but rather it is condensed on a molecular scale in the chromosomal DNA or in that of every other organelle in each cell. This "intelligence" is the *sine qua non* of life. Where does it come from? . . . This is a problem that concerns both biologists and philosophers, and, at present, science seems incapable of solving it. . . . If to determine the origin of information in a computer is not a false problem, why should the search for the information contained in cellular nuclei be one?[3]

Grassé argued that, due to their uncompromising commitment to materialism, the Darwinists who dominate evolutionary biology have failed to define properly the problem they were trying to solve. The real problem of evolution is to account for the origin of new genetic information, and it is not solved by providing illustrations of the acknowledged capacity of an existing genotype to vary within limits. Darwinists had imposed upon evolutionary theory the dogmatic proposition that variation and innovative evolution are the same process, and then had employed a systematic bias in the interpretation of evidence to support the dogma. Here are some representative judgments from Grassé's introductory chapter:

Through use and abuse of hidden postulates, of bold, often ill-founded extrapolations, a pseudoscience has been created. . . . Biochemists and biologists who adhere blindly to the Darwinist theory search for results that will be in agreement with their theories. . . . Assuming that the Darwinian hypothesis is correct, they interpret fossil data according to it; it is only logical that [the data] should confirm it; the premises imply the conclusions. . . . The deceit is sometimes unconscious, but not always, since some people, owing to their sectarianism, purposely overlook reality and refuse to acknowledge the inadequacies and the falsity of their beliefs.[4]

Dobzhansky's review succinctly summarized Grassé's central thesis:

The book of Pierre P. Grassé is a frontal attack on all kinds of "Darwinism." Its purpose is "to destroy the myth of evolution as a simple, understood, and explained phenomenon," and to show that evolution is a mystery about which little is, and perhaps can be, known.

Grassé was an evolutionist, but his dissent from Darwinism could hardly have been more radical if he had been a creationist. It is not merely that he built a detailed empirical case against the neo-Darwinian picture of evolution. At the philosophical level, he challenged the crucial doctrine of uniformitarianism, which holds that processes detectable by our present-day science were also responsible for the great transformations that occurred in the remote past. According to Grassé, evolving species acquire a new store of genetic information through "a phenomenon whose equivalent cannot be seen in the creatures living at the present time (either because it is not there or because we are unable to see it)."[5] Grassé acknowledged that such speculation "arouses the suspicions of many biologists . . . [because] it conjures up visions of the ghost of vitalism or of some mystical power which guides the destiny of living things. . . ." He defended himself from these charges by arguing that the evidence of genetics, zoology, and paleontology refutes the Darwinian theory that random mutation and natural selection were important sources of evolutionary innovation. Given the state of the empirical evidence, to acknowledge the existence of some as yet undiscovered orienting force that guided evolution was merely to face the facts. Grassé even turned the charges of mysticism against his opponents, commenting sarcastically that nothing could be more mystical than the Darwinian view that "nature acts blindly, unintelligently, but by an infinitely benevolent good fortune builds mechanisms so intricate that we have not even finished with analysis of their structure and have not the slightest insight of the physical principles and functioning of some of them."[6]

Dobzhansky disagreed with Grassé fundamentally, but he acknowledged at the outset that his French counterpart knew as much about the scientific evidence regarding animal evolution as anyone in the world. As he put it,

> Now one can disagree with Grassé but not ignore him. He is the most distinguished of French zoologists, the editor of the 28 volumes of *Traite de Zoologie*, author of numerous original investigations, and ex-president of the Academie des Sciences. His knowledge of the living world is encyclopedic.

In short, Grassé had not gone wrong due to ignorance. Then where *had* he gone wrong? According to Dobzhansky, the problem was that the most distinguished of French zoologists did not understand the rules of scientific reasoning. As Dobzhansky summed up the situation:

> The mutation-selection theory attempts, more or less successfully, to make the causes of evolution accessible to

reason. The postulate that the evolution is "oriented" by some unknown force explains nothing. This is not to say that the synthetic . . . theory has explained everything. Far from this, this theory opens to view a great field which needs investigation. Nothing is easier than to point out that this or that problem is unsolved and puzzling. But to reject what is known, and to appeal to some wonderful future discovery which may explain it all, is contrary to sound scientific method. The sentence with which Grassé ends his book is disturbing: "It is possible that in this domain biology, impotent, yields the floor to metaphysics."

I have begun with the Dobzhansky/Grassé exchange to make the point that whether one believes or disbelieves in Darwinism does not necessarily depend upon how much one knows about the facts of biology. Belief that the various types of plants and animals were created by an extension of the kind of changes Dobzhansky's experiments brought about in fruitflies, is at bottom a question of metaphysics. By metaphysics, I mean nothing more pretentious than the assumptions we all make about just which possibilities are worth considering seriously. For example, Pierre Grassé was willing to consider, and eventually to endorse, the possibility that the so-called "evolution in action" which the neo-Darwinists were observing is merely a variation or fluctuation that is not a source of evolutionary innovation. To put the point in the language used by some contemporary biologists, Grassé proposed to "decouple macroevolution from microevolution." Such proposals have generally floundered on the inability to establish sufficiently credible distinctive macroevolutionary mechanisms. (For example, the widely publicized "new theory" of punctuated equilibrium turned out to be just a gloss upon Ernst Mayr's thoroughly Darwinian theory of peripatric speciation.) What was different about Grassé was that he was willing to give unprejudiced consideration to the possibility that science does not know, and may never know, how new quantities of genetic information have come into the world.

From Dobzhansky's viewpoint, prejudice against such a possibility is a virtue, because to accept that kind of limitation would be to give up on science. As he saw it, we already know a lot about how plants and animal populations vary in the everyday world of ecological time. Dog breeders have given us St. Bernards and dachshunds, laboratory experiments have produced monstrous fruitflies, mainland species have differentiated after migrating to offshore islands, and the ratio of dark to light peppered moths in a population changed when the background trees were dark due to industrial air pollution. To be sure, none of these examples demonstrated the kind of innovation that Grassé had in mind. In the absence of a better theory, however, Darwinists consider it reasonable to assume that these variations illustrate the working in ecological time of a grand process that over geological ages created fruitflies and peppered moths and scientific observers in the first place. By making that extrapolation Darwinists create a scientific paradigm that can be fleshed out with further research, and improved. For a critic to suggest the possible existence of some factor outside the paradigm is helpful only if he or she can also propose a research strategy for investigating it. To Dobzhansky, therefore, Grassé's insistence that the sources of new genetic information might not be "accessible to reason" was pointless and harmful to the cause of science.

There is a political and religious dimension to the issues Grassé and Dobzhansky were debating, which must also be considered. To say as Grassé did that, in the domain of creation, "biology, impotent, yields the floor to metaphysics" is to imply something important about the relative cultural authority of biologists and metaphysicians. Whatever that might mean in France, in the United States the scientific establishment has been in conflict over evolution for generations with the advocates of creationism. Although the scientists have won all the legal battles, there are still a lot of creationists around who are very much unconvinced by what the Darwinists are telling them. How many there are depends upon how "creationism" is defined. The most visible creationists are the biblical fundamentalists who believe in a young earth and a creation in six, twenty-four hour days; Darwinists like to give the impression that opposition to what they call "evolution" is confined to this group. In a broader sense, however, a creationist is any person who believes that there is a Creator who brought about the existence of humans for a purpose. In this broad sense, the vast majority of Americans are creationists. According to a 1991 Gallup poll, 47 percent of a national sample agreed with the following statement: "God created mankind in pretty much our present form sometime within the last 10,000 years." Another 40 percent think that "Man has developed over millions of years from less advanced forms of life, but God guided this process, including man's creation." Only 9 percent of the sample said that they believed in biological evolution as a purposeless process not guided by God.

The evolutionary theory endorsed by the American scientific and educational establishment is of course the creed of the 9 percent, not the God-guided gradual creation of the 40 percent. Persons who endorse a God-guided process of evolution may think that they have reconciled religion and science, but this is an illusion produced by vague terminology. A representative Darwinist statement of "the meaning of evolution" may be found in George Gaylord Simpson's book bearing that title. In the words of Simpson:

> Although many details remain to be worked out, it is already evident that all the objective phenomena of the history of life can be explained by purely naturalistic or, in a proper sense of the sometimes abused word, materialistic factors. They are readily explicable on the basis of differential reproduction in populations (the main factor in the modern conception of natural selection) and of the mainly random interplay of the known processes of heredity. . . . Man is the result of a purposeless and natural process that did not have him in mind.[7]

The prestige of the scientific establishment, and of the intellectual class in general, is heavily committed to the proposition that evolution—as George Gaylord Simpson used the term—is either a fact, or a theory so well supported by evidence that only ignorant or thoroughly unreasonable people refuse to believe it.

If the scientists ever had to retreat on this issue, the cultural consequences could be significant. Persons who now have prestigious status as cultural authorities would be discredited, and the political and moral positions they have advocated might be discredited with them. That is the fear of Michael Ruse, author of *Darwinism Defended*. Ruse proclaims proudly that Darwinism reflects "a strong ideology," and "one to be proud of." According to Ruse, contemporary Darwinians "show a strong liberal commitment" in both their politics and their sexual morality.[8] Advocates of creation, on the other hand, want to restore a "morality based on narrow Biblical lines" with respect to marriage and sexual behavior. Upholding Darwinism is therefore an important way of protecting political liberalism, feminism, and the sexual revolution of the 1960s. Ruse concludes his book with these stirring lines "Darwinism has a great past. Let us work to see that it has an even greater future."[9]

Such statements are equivalent to the claims of creation-science advocates that to doubt the Genesis account is to open the floodgates for all kinds of immorality. I think that Michael Ruse and Henry Morris are both right to insist that cultural acceptance of Darwinism has important consequences for politics and morality. Recognition of this factor, however, also has important implications for how we should regard Darwinism's rules of reasoning. Are those rules designed to protect a charter of liberty from scientific criticism—criticism that might, wittingly or unwittingly, give aid and comfort to persons who want to deprive the Darwinist establishment of its cultural authority? If physicists were to start to proclaim that the Big Bang has had a wonderful past, and we must all work to see that it has a wonderful future, I am sure we would all lose confidence in their ability to assess objectively the arguments of Big Bang critics.

Darwinism's rules of reasoning not only protect the cultural authority of Darwinists. They also permit Darwinist writers to take the mutation/selection paradigm for granted even when they are describing evidence that directly contradicts it. This feat of intellectual contortionism is strikingly illustrated by Stephen Jay Gould's book, *Wonderful Life*. Gould's best seller adds a great deal to our knowledge of the "Cambrian explosion," meaning the sudden appearance of the invertebrate animal phyla, without visible ancestors, in the 600 million-year-old rocks of the Cambrian era. Unicellular life had existed for a long time, and some multicellular groups appear in the immediately Precambrian rocks, but nothing can be established as ancestral to the Cambrian animals. As Richard Dawkins described the situation, "It is as though [the Cambrian phyla] were just planted there, without any evolutionary history."[10]

In recent years the mystery has deepened, because it appears that the Cambrian animal groups were far more varied than had been imagined. The more distinct groups that there were in the Cambrian, the more chains of ancestors there ought to have been in the Precambrian. Some remarkable Cambrian fossils found in a Canadian formation known as the Burgess Shale were originally classified in familiar groups. Gould explains that the discoverer of the Burgess Shale fossils, Charles Walcott, tried to "shoehorn" the odd creatures into familiar taxonomic categories because of his predisposition to avoid multiplying the difficulties of what is called the "artifact theory" of the Precambrian fossil record. As Gould explains the problem:

> Two different kinds of explanations for the absence of Precambrian ancestors have been debated for more than a century: the artifact theory (they did exist, but the fossil record hasn't preserved them), and the fast-transition theory (they really didn't exist, at least as complex invertebrates easily linked to their descendants, and the evolution of modern anatomical plans

occurred with a rapidity that threatens our usual ideas about the stately pace of evolutionary change).

Reclassification of the Burgess Shale fossils has now established some fifteen or twenty species that cannot be related to any known group and therefore constitute distinct and previously unknown phyla. There are also many other species that can fit within an existing phylum but are still remarkably distinct from anything known to exist earlier or later. The general history of animal life is thus a burst of general body plans followed by extinction. Many species exist today which are absent from the rocks of the remote past, but they fit within general taxonomic categories present from the very beginning. Darwinian theory predicts a "cone of increasing diversity," as the first living organism, or first animal species, gradually and continually diversified to create the higher levels of the taxonomic order. The animal fossil record more resembles such a cone turned upside down, with the phyla present at the start and thereafter decreasing. In short, the more we learn about the Cambrian fossils, the more difficult it becomes to see them as the product of Darwinian evolution.

Gould describes the reclassification of the Burgess fossils as the "death knell of the artifact theory," because it adds so many new groups that appear without Precambrian ancestors.

> If evolution could produce ten new Cambrian phyla and then wipe them out just as quickly, then what about the surviving Cambrian groups? Why should they have had a long and honorable Precambrian pedigree? Why should they not have originated just before the Cambrian, as the fossil record, read literally, seems to indicate, and as the fast-transition theory proposes?[11]

A mysterious process that produces dozens of complex animal groups directly from single-celled predecessors, with only some words like "fast-transition" in between, may be called "evolution"—but the term is being used more in the sense of Grassé's heresy than of Dobzhansky's Darwinian orthodoxy. Each of those Cambrian animals contained a variety of immensely complicated organ systems. How can such innovations appear except by the gradual accumulation of micro-mutations, unless there was some supernatural intervention? It is not only that the Darwinian theory requires a very gradual line of descent from each Cambrian animal group back to its hypothetical single-celled ancestor. Because Darwinian evolution is a purposeless, chance-driven process, which would not proceed directly from a starting point to a destination, there should also be thick bushes of side branches in each line. As Darwin himself put it, if Darwinism is true the Precambrian world must have "swarmed with living creatures" many of which were ancestral to the Cambrian animals. If he really rejects the artifact theory of the Precambrian fossil record, Gould also rejects the Darwinian theory of evolution.[12]

Readers familiar with Gould's writings know that he has at times expressed great skepticism concerning the neo-Darwinian theory that Dobzhansky proclaimed so confidently. In a paper published in *Paleobiology* in 1980, Gould wrote that, although he had been "beguiled" by the unifying power of neo-Darwinism when he studied it as a graduate student in the 1960s, the weight of the evidence has since driven him to the reluctant conclusion that neo-Darwinism "as a general proposition, is effectively dead, despite its persistence as textbook orthodoxy."[13] In place of the dead orthodoxy Gould predicted the emergence of a new macroevolutionary theory based on the views of geneticist Richard Goldschmidt, another heretic whose views were every bit as obnoxious to Darwinists as those of Grassé. The new theory did not arrive as predicted, however, and Gould subsequently seems to have heeded Dobzhansky's admonition: if you can't improve on the mutation/selection mechanism, don't trash it in public.

For whatever reason, Gould did not point out to his readers that the utterly un-Darwinian Cambrian fossil record provides no support whatever for claims about the role of mutation and selection in the creation of complex animal life, or for metaphysical speculations about the purposelessness of the process that created humans. Instead, he indulged freely in just such speculation himself, rightly judging that his audience of intellectuals would accept uncritically his casual assumption of metaphysical naturalism. In the concluding chapter he commented on a Burgess Shale fossil called *Pikaia*. Walcott classified *Pikaia* as a worm, but a more recent study concludes that the creature was a member of the phylum Chordata, which includes the subphylum Vertebrata, which includes us. That for Gould means that *Pikaia* might be our ancestor, which implies that, unlike many other Burgess Shale creatures, it left descendants. If *Pikaia* had not survived the mass extinctions that killed off so many other Cambrian fossil creatures, we would never have evolved. The existence of humans is therefore not a predictable consequence of evolution, but a never-to-be-repeated accident. Gould concluded this reflection, and the book, with the following sentence:

> We are the offspring of history, and must establish our own paths in this most diverse and interesting of conceivable universes—one indifferent to our suffering, and therefore offering us maximum freedom to thrive, or to fail, in our own chosen way.

Of course absolutely nothing in the Burgess Shale fossils supports Gould's speculation that the universe is indifferent to our sufferings, or discredits the belief that we are responsible to a divine Creator who actively intervened in nature to bring about our existence. On the contrary, the genuine scientific portion of *Wonderful Life* provides ample grounds for doubting the expansive notions of metaphysical naturalists like Theodosius Dobzhansky and George Gaylord Simpson. But because of Darwinism's rules of reasoning, even anti-Darwinian evidence supports Darwinism.

The statement defining the agenda for this symposium asserts that an *a priori* commitment to metaphysical naturalism is necessary to support Darwinism. *Methodological* naturalism—the principle that science can study only the things that are accessible to its instruments and techniques—is not in question. Of course science can study only what science can study. Methodological naturalism becomes metaphysical naturalism only when the limitations of science are taken to be limitations upon reality. If the history of life can involve only those natural and material processes that our science can observe, then either Darwinism or something very much like it simply must be true as a matter of philosophical deduction, regardless of how scanty the evidence may be. Add to this the requirement that critics of a paradigm must propose an alternative—and we have the metaphysical rules of Dobzhansky.

I do not doubt that Darwinian evolution will continue as the reigning paradigm as long as Dobzhansky's metaphysical rules are enforced. To say this is merely to say that the neo-Darwinian synthesis is the most plausible naturalistic and materialistic theory for the development of complex life that is now available. That proposition in turn is virtually a tautology, because the synthesis is a vague and flexible conglomeration that readily incorporates any seemingly non-Darwinian elements—such as the molecular clock or punctuated equilibrium—that appear from time to time.[14] If Dobzhansky makes the rules, Darwinism wins; but what happens if we evaluate the theory by Pierre Grassé's rules? I have argued my position on the evidence at book length in *Darwin on Trial*, and I will not go over that ground again now. My concern on this occasion is merely to speak about how we can conduct a fair and illuminating discussion of this subject.

I propose that we avoid using the word *evolution* altogether, or at least that we carefully specify what meaning we have in mind when we do use the term. The problem is that "evolution" has many meanings, some of which are controversial and some of which are not. Nobody, including the creation-scientists, denies that selection by human intelligence can cause a degree of variation, of the kind seen in the breeding of domestic animals or fruitflies. Nobody denies that mutation and selection have caused variation in nature, as with the varieties of shapes and colors in the famous finches of the Galápagos Islands or the shifting ratios of dark and light peppered moths in the midlands of England. As we have seen, Pierre Grassé denied that these observations illustrate "evolution," because they merely bring out the capacity for variation in an existing genotype and do not involve the introduction of new genetic information.

If we are going to discuss this argument, it can only confuse matters to make statements like "The evidence of biogeography provides ample evidence of evolution." Of course it does, but does it illustrate the kind of evolution that nobody disputes or the kind that many of us, including eminent biologists, *do* dispute? Biogeography does tell us that certain marsupial mammals exist only in Australia, for example. What else does it tell us about the process that created them?

I have found it helpful when discussing Darwinism to speak not of "evolution" but rather of the "blind watchmaker thesis," after the title of the famous book by Richard Dawkins. This book is the outstanding contemporary defense of the part of Darwinism that is really interesting: the claim that natural selection can accomplish wonders of creation, and not merely a degree of diversification. According to Dawkins, "Biology is the study of complicated things that give the appearance of having been designed for a purpose."[15] This is essentially what Pierre Grassé had in mind when he compared living organisms to things like cathedrals and computer programs that are designed by human intelligence for a purpose. Of course, Dawkins argues that this appearance is misleading, because the features that appear to have been designed were in fact produced by the purposeless, unintelligent processes of mutation and selection.

Whether this argument is supported by evidence when it is considered without prejudice is the fundamental point at issue. Prejudice enters the discussion if, for example, we define "science" as requiring an *a priori* assumption of metaphysical naturalism. In that case, the blind watchmaker thesis simply has to be true as a matter of philosophical deduction, and the scientific evidence is relevant only to illustrate a doctrine that we know to be true in advance.

My first proposal is that we should define terms carefully and use them consistently, trying at all times to illuminate points of disagreement rather than to dismiss them with semantic devices, such as the use of argumentative definitions of "evolution" or "science." My second point is that we should give careful consideration to the appropriate role of theological arguments in scientific discussions of Darwinism. I am referring here not to those creationists who invoke the Bible, but to the important role that a theological argument—"God wouldn't have done it this way"—plays in Darwinist apologetics. For example, Stephen Jay Gould's famous argument in *The Panda's Thumb* takes this form: A proper Creator would not have made the Panda's thumb from a wristbone, or used homologous components in orchids. To quote Gould:

> Orchids manufacture their intricate devices from the common components of ordinary flowers, parts usually fitted for very different functions. If God had designed a beautiful machine to reflect his wisdom and power, surely he would not have used a collection of parts generally fashioned for other purposes. Orchids were not made by an ideal engineer; they are jury-rigged from a limited set of available components. Thus, they must have evolved from ordinary flowers.[16]

And of course "evolution" implies the blind watchmaker thesis, which implies that we live in a purposeless cosmos that cares nothing for our sufferings. David Hull makes a similar argument in his review for *Nature* of *Darwin on Trial*. On the time-honored theory that the best defense is a good offense, Hull defends the blind watchmaker thesis by attacking the divine creation alternative. The world is full of waste and cruelty: therefore God didn't create it and therefore the blind watchmaker presumably did. I could leave the matter there, but I enjoyed Hull's chamber of horrors so much that I will quote the relevant passage:

> What kind of God can one infer from the sort of phenomena epitomized by the species on Darwin's Galápagos Islands? The

17

evolutionary process is rife with happenstance, contingency, incredible waste, death, pain and horror. Millions of sperm and ova are produced that never unite to form a zygote. Of the millions of zygotes that are produced, only a few ever reach maturity. On current estimates, 95 per cent of the DNA that an organism contains has no function. Certain organic systems are marvels of engineering; others are little more than contraptions. When the eggs that cuckoos lay in the nests of other birds hatch, the cuckoo chick proceeds to push the eggs of its foster parents out of the nest. The queens of a particular species of parasitic ant have only one remarkable adaptation, a serrated appendage which they use to saw off the head of the host queen. . . . Whatever the God implied by evolutionary theory and the data of natural history may be like, He is not the Protestant God of waste not, want not. He is also not a loving God who cares about His productions. He is not even the awful God portrayed in the book of Job. The God of the Galápagos is careless, wasteful, indifferent, almost diabolical. He is certainly not the sort of God to whom anyone would be inclined to pray.

Simpson tells us that the world is purposeless because Darwinian evolution did all the creating. Gould and Hull tell us that Darwinian evolution must have done the creating because the characteristics of organisms imply a world devoid of purpose. A wise and benevolent creator would not employ homologous parts; would not waste millions of sperm and ova when one pair would suffice; would not countenance the deplorable ethics of the cuckoo; and would not even allow the variations in finches and turtles that Darwin observed in the Galápagos. These particular examples don't seem persuasive to me, but lurking behind them is the well-known argument from evil and undeserved suffering that forms the background to some of the world's greatest literature, from the book of Job to *Paradise Lost* to *The Brothers Karamazov*. Yes, the world is full of waste and suffering, and also nobility and beauty. If that is all that is necessary to establish Darwinian evolution, then Darwinian evolution is established. But do we call this kind of reasoning *science*?

I am not going to address the philosophical arguments against theism on this occasion, because my position is that speculation about what God would or would not have done should play no part in scientific discussion. If others want to bring theology into the picture, that is fine with me, but I want them to recognize that the will of God is not a subject over which biologists have professional jurisdiction. If we are going to debate theology the theologians are going to have a place at the table, and that includes creationist theologians. If Darwinists want to avoid the situation predicted by Grassé, where biology yields to metaphysics, I suggest that they agree to put theological speculations aside.

Leaving theology out of the discussion doesn't mean that scientists should assume confidently that God does not exist and go on to build philosophical theories on that foundation. What it does mean is that scientists should try to find out as much as they can about how the world works through empirical investigation, recognizing that an appropriately humble science may be unable to come to confident conclusions about matters that are difficult to observe. Science should be more than just a weapon that metaphysical naturalists wield in their arguments with theists. It should be a self-critical search for as much of the truth as its methods of investigation can ascertain, which may or may not include the truth about how new quantities of genetic information have come into the world.

NOTES

[1]Pierre P. Grassé, *L'Evolution du Vivant* (1973), published in English translation as *The Evolution of Living Organisms* (1977) (hereafter Grassé). The review of the original French edition by Dobzhansky, titled "Darwinian or 'Oriented' Evolution?" appeared in *Evolution*, vol. 29 (June 1975), pp. 376-378.

[2]Grassé, p. 130.

[3]Grassé, p. 2.

[4]Grassé, pp. 7-8.

[5]Grassé, p. 208. See also p. 71: "We are certain that it [evolution] does not operate today as it did in the remote past. Something has changed. . . . The structural plans no longer undergo complete reorganization; novelties are no longer plentiful. Evolution, after its last enormous effort to form the mammalian orders and man, seems to be out of breath and drowsing off."

[6]Grassé, p. 168.

[7]George Gaylord Simpson, *The Meaning of Evolution* (rev. ed., 1967), pp. 344-345.

[8]Michael Ruse, *Darwinism Defended* (Addison-Wesley, 1982), p. 280.

[9]Ruse, pp. 328-329.

[10]Richard Dawkins, *The Blind Watchmaker* (Longman, UK, 1986), p. 229.

[11]Stephen Jay Gould, *Wonderful Life* (1989), pp. 271-273.

[12]Careful readers will note that the non-existence of the Cambrian ancestors is vaguely qualified by the phrase "at least as complex invertebrates easily linked to their descendants." I have learned to be alert to this sort of qualification in Gould's writing, because it signals a possible line of retreat. I have reason to believe that Gould would repopulate the Precambrian world with invisible ancestors, and thus re-embrace the artifact theory, if he were accused of abandoning the mutation/selection mechanism and thus leaving unexplained the evolution of complexity.

[13]Stephen Jay Gould, "Is a New and General Theory of Evolution Emerging?" *Paleobiology*, vol. 6 (1980), pp. 119-130. Reprinted in the collection *Evolution Now: A Century After Darwin* (Maynard Smith, ed., 1982).

[14]Stephen Jay Gould has complained that vagueness in the definition of the neo-Darwinian synthesis "imposes a great frustration upon anyone who would characterize the modern synthesis in order to criticize it." Gould, "Is a New and General Theory of Evolution Emerging?" pp. 130-131, in the collection *Evolution Now: A Century After Darwin* (Maynard Smith, ed., 1982).

[15]Dawkins, p. 1.

[16]Stephen Jay Gould, *The Panda's Thumb*, p. 20.

2
Darwinism:
Philosophical Preference, Scientific Inference, and Good Research Strategy
Michael Ruse

IN 1859, CHARLES DARWIN published his great work, *On the Origin of Species*. He claimed that all organisms, including ourselves, are the products of a slow, natural process of development—"evolution"—from just one or a few forms. As you might imagine, much that Darwin had to say, has been revised in the course of a century and a half of research. But I think that in essence Darwin was right. In this paper I shall defend my beliefs. I am interested here only in the positive case, and therefore shall have nothing to say of a negative nature about those who do not share my beliefs. I am sure they can speak for themselves, and I welcome the opportunity to let them.

I shall divide my discussion into three parts: that dealing with the underlying philosophical commitment to science; that dealing with the fact of evolution and that dealing with the belief that the right scientific strategy is the Darwinian one, referring here to Darwin's major mechanism of "natural selection."[1]

The Commitment to Science

I am not a scientist, but I believe that the proper and most profitable way to explore and understand this wonderful world of ours is the *scientific* one. I am not implying that scientists are better or worse people than the rest of us, but I do think that their methodology is the best one. By "scientific methodology" or "attitude" in this case, I mean a commitment to the idea of the world being law-bound—that is, subject to unbroken regularity—and to the belief that there are no powers, seen or unseen, that interfere with or otherwise make inexplicable the normal workings of material objects. I am not trying here for a trick definition, and I include such things as gravity and electricity in the material world. I recognize, and am thrilled at the fact, that the world is full of mysterious things like electrons; it is just that they are not so mysterious as to lie outside law.

I take it that my position excludes certain sorts of miracles—for instance, Jesus turning the water into wine (taken in a literal sense). On the other hand, if your miracle does not interfere with the workings of nature, and if you do not make any scientific claims for your beliefs, I have nothing to say, *qua* my commitment to science. For instance, I see no reason why one should not be a scientist in my sense and also believe in the Catholic doctrine of transubstantiation, the belief that the water and wine turn into the body and blood of Christ during the Catholic mass. No priest would ever claim that you cut open the loaf and the flesh oozes out. The Aristotelian distinction between substance and accident lies beyond science, as I am understanding it.

Is my position reasonable, provable, irrational, or just a philosophical preference? It is certainly not provable in the sense that the theorem of Pythagoras is provable. It is not provable in the sense that one can prove that the earth goes around the sun; it is the presupposition of that particular claim! On the other hand, I deny that it is merely an irrational prejudice, or even "just" a philosophical preference. Although you must remember that I am prejudiced here, because I am a professional philosopher.

Sometimes you have beliefs which you cannot prove absolutely, but which are still reasonable in the sense that you can offer good arguments. Political beliefs probably fall into this category—but let me take the less contentious (or is it more contentious?) case of preferences about sport. Is baseball a better game than cricket? Ultimately, there is no way of deciding; some people like cricket and some people like baseball. But you can offer some reasons for your preference. For instance, a strong argument in favor of baseball is that the game gets finished. You do not go four-and-a-half days, have the game in the balance, and then get everything washed out, literally, by the English weather.

In the same way, you can offer an argument for the reasonableness of the scientific attitude. A great many things that in the past were thought miraculous are now seen to be covered by law. To take but one example, it was thought by the well-to-do in New York City in the last century that typhoid was a miraculous punishment by God of the Irish and other poor immigrants for their dissolute lifestyles and stubborn adherence to Popery. Now we know that only the bottom segment of society had to drink the stinking, contaminated water. There was no miracle about that, or about the diseases the lowest group-members thereby contracted. Of course, you can always say that there was no miracle then, but there *is* a genuine miracle coming up around the next corner. I say that it is not reasonable to believe this. The mature attitude is to go with unbroken law.

I should say one final thing before I move on. I recognize that having a commitment to law is one thing. It is another thing to put such a commitment into effect. I do not want to write an essay here about scientific method, but obviously what I am presuming is that the working scientist will be testing his/her hypotheses against the world of experience (which may well include the manipulated world of experience of the experimenter). Also, to guide him/her, the scientist will be relying on well-established rules of scientific method: the appeal to consistency, the preference for simplicity, the aim for comprehensiveness.

This last dictate I take to be very important indeed, especially in the kind of case we are about to consider here. Scientists aim to include all of their subsidiary hypotheses beneath one or two major all-embracing laws. The paradigmatic example of such a "consilience of inductions" was Isaac Newton's showing that the terrestrial mechanics of Galileo and the heavenly dynamics of Kepler could be subsumed beneath shared laws of motion and of gravitational attraction.

The status and proof of such rules of methodology have been matters of some debate. I myself favor an explanation that roots the rules in our evolutionary past. But that is not a significant issue here. I agree that the rules are not provable absolutely in logic; but, as with the general commitment to lawfulness, I argue that their acceptance is more than a matter of arbitrary whim. Return for a moment to the baseball analogy, although I accept that there is an element of subjectivity about the baseball case that is missing in the case of science. Even though it is true that you cannot prove absolutely that some rules of baseball are necessary, through long experience you know that unless you have rules, and rules of a certain kind, your game will cease to function in the way that you expect of a top-notch sport. If, say, a runner did not have to touch all of the bases on his way round, baseball simply would not work. The same is true of science with respect to simplicity and consilience and the like. Science will not work without rules, and experience tells us which are the best rules.

The Fact of Evolution
Grant now that one is going to accept the scientific way of thinking.

The next question is, What should we believe about origins, organic origins (including ourselves)? We have a range of options. Logically, it could be the case that life is as old as the universe, and that it has always existed in the present form. This includes the possibility that the universe is eternal. Or, it could be that life comes into being spontaneously, on a regular basis, a kind of organic equivalent to the steady-state universe. It could even be that limbs and the rest form by pure chance, out of randomly moving molecules, and then that these sometimes join together to make functioning organisms. This was the belief of the Greek atomists, who reasonably thought that if one had infinite space and infinite time, anything might happen!

I myself do not believe any of these things. Everything I know (although, I admit candidly that I get it all at secondhand) suggests that the universe formed about twenty billion years ago, in a big bang, and that everything has been expanding and evolving ever since. But, remembering now that I take the scientific attitude for granted, why go on to accept organic evolution? Why accept that all organisms came about through a natural process of development? Why believe that we humans had monkeys for grandfathers?

If challenged with this point, many people, especially those who are not professional evolutionists, would say, "Because of the fossils." They would think that the record of the rocks is the ultimate ground for belief. Along with professional evolutionists, however, I would cast my net much wider. If you want to find out about life's history, the fossil record is invaluable. If you want to find out what our ancestors were like, then you must go and dig up the bones in Africa. Today we have many other techniques for inferring organic histories ("phylogenies"), often techniques at the molecular level, but the fossils remain crucial.

I am an evolutionist as such, because of all of the evidence. I find particularly convincing the evidence of morphology. Why is it that the limbs of vertebrates, used for all sort of different purposes, have the same isomorphic pattern of bones ("homologies")? Why do we find repetition between the forelimb of the human (a grasping instrument), the front leg of the horse (running), the flipper of the whale (swimming), the wing of the bat (flying), and more? My answer is that if you think in terms of unbroken law, then evolution makes the most sense.

Like Darwin himself, I also find very impressive the facts of biogeographical distribution. That famous group of islands in the

Pacific, the Galápagos Archipelago, has different species of finch from island to island. How could this be, other than through the gradual development from shared ancestors? If you think of seeding from outer space, or of spontaneous generation, or some such thing, there is no reason why the finches appear thus so close together. Why not one species of finch in the Galápagos and another in the Hebrides?

Am I making this claim about evolution as a matter of absolute logical necessity, given the commitment to science? The phrase I like is that of the lawyers: "Beyond reasonable doubt." I think that the fact of evolution is beyond reasonable doubt. There is no need for the student of biology to take seriously, say, the hypothesis of spontaneous generation (of whole forms). Ideas like that have been considered and discredited. You will recognize that here I am appealing to a consilience of inductions. My claim is that evolution brings many disparate parts of biological science together and unites them beneath one all-embracing hypothesis. It is not reasonable to go on questioning.

Continuing the legal analogy, you will realize that this is precisely what we do in a court of law, especially when we are dealing with circumstantial evidence. The guilt for the murder is pinned on the butler, say, because the fingerprints and bloodstains and the broken alibi and the motive and everything else all make sense on the hypothesis of the butler's guilt. Likewise, evolution is a reasonable belief because the homologies and the biogeographical distributions and all the rest make sense on the hypothesis of the truth of evolution. The butler's guilt and the truth of evolution are "beyond reasonable doubt."

The truth of evolution is not a logical necessity; as in law, new evidence could lead one to reconsider even a verdict decided as "beyond reasonable doubt." But it is going to take a lot of evidence of a very strong nature. I am not holding my breath in anticipation. I would put the chances of my being wrong on this point about on a par with my favorite tabloid being right that Elvis is indeed alive and well and living in retirement in Florida.

Darwinism as the Best Scientific Strategy

We come now to the question of theories or causes. Supposing that one is an evolutionist, what drives the process? There have been lots of suggested mechanisms: natural selection, Lamarckism (the inheritance of acquired characters), saltationism (jumps), orthogenesis (directed trends), drift (random meandering), and more. True to my intentions, I am neither going to list all of the options nor refute them. I am concerned with my position, Darwinism, and why I think that it is the right choice.

My answer centers on what I think is the most important aspect of organic nature, separating it from the rest of reality. Organisms work, they function, they are as if designed; they have adaptations. It makes no sense to ask about the function of the craters on the moon. It makes every sense to ask about the function of the sail on the back of *Dimetrodon*. I have said that organisms are "as if" designed. Am I suggesting that they are not designed? No, nor am I saying that they are designed. My point is simply that, as one following science, if talk of a designer implies someone who got involved miraculously in the process, that idea is simply inappropriate in this context.

One must explain the adaptedness of organisms by natural means. This one can do by invoking natural selection, the differential survival of certain organisms, against a background of modern genetics. Originally this was Mendelian genetics, but now we think that ultimately all can be referred back to long chains of ribonucleic acid (DNA and RNA). I believe that, for reasons not connected with their needs, organisms vary. There is a constant pressure of population, leading to competition for resources and mates ("the struggle for existence"). Some organisms survive and reproduce. Some do not. Those that do succeed, have on average variations not possessed by the losers, and those variations, although they appeared for no good reason, helped the winners to win. Given enough time, since the variations are heritable, one gets evolution. More significant, one gets the evolution of adaptations.

The argument is simple, but not simplistic. It is true, but not a truism. It is one of the most beautiful and powerful discoveries ever made by a human being. That is why I am proud to be called a "Darwinian." It is not just that natural selection explains the world that we know already, but that like the very best scientific ideas, it contains potential for explanation in new areas of inquiry. In the last thirty years, for instance, Darwin's selection has been introduced into our study of social behavior in a major and satisfying manner. Such old problems, as for example, the structure of nests of the hymenoptera (ants, bees, and wasps), are now seen in an altogether new and revealing light.

Does this mean that I believe that natural selection is the exclusive cause of evolutionary change? If I did, I would be the first person ever to do so—and I am not about to set any records here! I see no logical reason why every last aspect of the organic world should be adapted, and I very much doubt that it is. Perhaps it is the case that male nipples have some adaptive function—but I rather doubt it. Even the sexual argument does not convince. If you can learn to love people without foreskins, then I see no reason why you should not learn to love people without nipples.

It seems to me quite plausible that there are reasons of change other than direct selective advantage: correlation with other, desirable features; by-production from changes of unrelated organs; and just plain chance. This latter might be very important at the molecular level, although do not ask me how important. I am sure that natural selection is very important; I am sure that it is the most important. I just doubt that it is the exclusive cause of evolutionary change.

I have spoken in my heading to this section of Darwinism as a "strategy" and that is precisely what I mean. I think that it is true. More important, I know that it works. It explains the world we know, and it lead us into new realms of the world that we do not yet know. It makes sense as a scientist, as an evolutionist, to back it for all that it is worth. That is what sensible strategies are all about.

Conclusion

As promised, I have simply stated what I believe. Let me say one final thing. I have offered my ideas in a true state of reverence. You may not agree with them. That is your right, and we can now start a debate. What I do say now, and shall always maintain, is that—whether or not there is some meaning to life above and beyond us all—great creations of the human spirit in themselves confer a meaning and significance to human existence. They dignify us, just as obscenities like the Holocaust degrade us. Even if all were destroyed tomorrow, nothing could negate the nobility of Plato's *Republic*, the beauty of Mozart's *Don Giovanni*, or the heady excitement of Darwin's hypothesis of organic evolution through natural selection. I am proud to be a human.

NOTES

[1]Since this is the nature of a personal essay, with a total lack of modesty I shall refer the reader only to books I have myself written or edited. For a general background to the logic of evolutionary thought, see *The Philosophy of Biology* (London: Hutchinson, 1973). Although now somewhat dated, it covers the main points in a fairly thorough

way. For a general background to Darwin's achievements, see *The Darwinian Revolution* (Chicago: University of Chicago Press, 1979). A vigorous defense of Darwinism can be found in *Darwinism Defended* (Boston: Addison-Wesley, 1982). My own philosophy of evolution, looking both at epistemology and ethics is in *Taking Darwin Seriously* (Oxford: Blackwell, 1986). Arguments for and against evolutionism and Creationism, concentrating on philosophical questions can be found in the edited collection *But Is It Science?* (Buffalo: Prometheus, 1988). And if you are still reading after all of this, you can find many references to the logic of biology in my handbook, *Philosophy of Biology Today* (Albany: SUNY Press).

3
Laws, Causes, and Facts
Response to Michael Ruse
Stephen C. Meyer

I APPRECIATE VERY MUCH the opportunity to respond to Professor Ruse. Though it is in the nature of a response to disagree, I must say that I always appreciate his work in philosophy of biology. His arguments are presented forcefully and cogently. Because of this they have always served to clarify my thinking even when I find myself on the opposite side of a particular philosophical issue.

As a philosopher of science, I also appreciate the title of Professor Ruse's paper; I too believe that an idea can be both a philosophical preference and a scientific inference. Recent work in philosophy of science on something called the "demarcation problem" suggests that it may be difficult to separate philosophical and scientific considerations in part because both science and philosophy share a common concern with explanation. Establishing a rigid line of demarcation between science and philosophy is especially difficult in the vexing world of origins research. So I appreciate Professor Ruse's drawing our attention to what is perhaps a false dichotomy in the title of this conference, the one between "scientific inference" and "philosophical preference."

Nevertheless, my philosophical preferences are somewhat different from Professor Ruse's. It will be part of the purpose of this response to suggest that inferences with decidedly theistic implications may also be considered properly scientific. In fact, I would like to suggest that although Professor Ruse's definition of science certainly serves certain philosophical preferences, it does not always promote theoretical openness, nor as a result, intellectual rigor.

In this response, I want to challenge two assertions that Professor Ruse has made. The first concerns his definition of the scientific attitude. Second, I want to challenge his claim that evolution, defined as common descent,[1] is a fact.

Challenge to Ruse's Definition of Science
Professor Ruse has suggested that to adopt the scientific outlook one

must accept that the universe is subject to natural law,[2] and that further, one must never appeal to (an intervening) agency as an explanation for events. Instead we must always look to what he calls "unbroken law" if we wish to explain things as scientists.

There are several problems with this assertion and with the so-called "covering law" conception of science that underlies it and to which Professor Ruse has appealed.[3] Indeed, unsolved problems with the covering law idea of science are legion.[4] It can be no purpose of mine, however, to recall or explain all of them. Nevertheless, one of the more salient difficulties with this philosophy of science—i.e., this theory about what constitutes a proper scientific theory—is relevant to my critique of Professor Ruse's suggestion that science is primarily concerned with explanation via natural law.

This difficulty is as follows: *the covering law model incorrectly conflates scientific laws and explanations*. There are two sides to this difficulty.

1. In the first place, many laws are descriptive and not explanatory. Many laws describe regularities, but do not explain why the events they describe occur. A good example of this from the history of science was Newton's Universal Law of Gravitation, which Newton himself freely admitted did not explain, but instead merely described, gravitational motion. As he put it in the "General Scholium" of the second edition of the *Principia: "Hypothesis non fingo"* (i.e., "I do not feign hypotheses").[5] To assert that science must explain by reference to "natural law" would necessarily eliminate from the domain of the properly scientific all fundamental laws of physics that describe mathematically, but do not explain, the phenomena they "cover."

2. Laws cannot be equated with causes or explanations for a second reason. Many scientific explanations do not depend, either principally, or at all, upon scientific laws. Many scientific explanations depend primarily upon antecedent causal conditions and events, not laws, to do what I have called the "primary explanatory work."[6] That is to say, citing past causal events often does more to explain a particular phenomenon than citing the existence of a regularity in nature. This is, in part, because many things do not come into existence via a series of events that regularly reoccur. For example, if a historical geologist seeks to explain the unusual height of the Himalayas, he or she will cite particular antecedent factors that were present in the case of the Himalayan orogeny but were not present in other mountain-building

episodes. Knowing the laws of geophysics that describe orogeny generally (if there even are such things) will aid the geologist very little in accounting for the contrast between the Himalayan and other orogenies. What the geologist needs in this situation for an explanation is not knowledge of a general law, but evidence of a particularly distinctive set of past conditions.[7]

The situation is similar to the situation faced by historians generally. Historical explanations of why World War I began—whether it was the ambition of the Kaiser's generals, the Franco-Russian defense pact, or the assassination of Archduke Franz Ferdinand—invariably and primarily involve the citation of *events*, conditions, or actions (and not laws) that are believed to be causally antecedent to the *explanandum*. As Michael Scriven has shown, we can often know what caused something (both in history and in disciplines like historical geology) even if we cannot relate causes and effects to each other as formal nomological statements.[8] Similarly, William Alston has shown that laws alone often do not explain particular events even when we have them.[9] Thus he concludes that to equate a law with an explanation or cause "is to commit a 'category mistake' of the most flagrant sort."[10]

Perhaps another example will help. If I wish to explain why human beings were able to fly to the moon, when apples usually fall to the earth, I will not primarily cite the law of gravity. Such a law is far too general to be primarily relevant to explanation in this context, because the law allows for a vast array of possible outcomes depending on initial and boundary conditions. The law stating that all matter gravitates according to an inverse-square law is consistent both with an apple falling to the earth and with an astronaut flying to the moon. Explaining why the astronaut flew, when apples routinely fall, therefore, requires more than citing the law, because the law is presumed operative in both situations. Accounting for the differing outcomes (i.e., between the apple and the astronaut) will require references to the antecedent (prior) conditions and events, which differed in the two situations. In other words, explanation in this case requires an accounting of the way in which engineers have altered the initial and boundary conditions affecting the astronauts to allow them to overcome the constraints ordinarily imposed by gravity on all earthbound objects.

Confusion about the role of antecedent conditions and laws in scientific explanation has led many to create a false dichotomy between "unbroken law" and the action of agency. In Professor Ruse's case this dichotomy is manifest in his assertion that invoking the action of a divine agent constitutes a departure from a commitment to natural law. I disagree. Posing the action of agents against the laws of nature creates an unnecessary dichotomy. The reason for this is simple. Agents can change initial and boundary conditions, yet in so doing they do not violate laws. Most scientific laws have the form, "If A then B will follow, given conditions X." If X are altered, or if A did not obtain, then it constitutes no violation of the laws of nature to say that B did not occur, even if we expected it to. Agents may alter the course of events, or produce novel events that violate our expectations, without violating the laws of nature. To assert otherwise is to misunderstand the distinction between antecedent conditions and laws.

The tendency to conflate laws with causes, on the one hand, and to treat natural laws and agency as mutually exclusive ontologies, on the other, has produced a confused set of expectations about what scientifically acceptable origins-theories must look like. This confusion has been heightened by the positivist tendency to see all scientific practice as identical and by talk about *the* scientific method.

In my own research, I have argued that a clear and logical distinction exists between questions that motivate historical and nonhistorical (or what might be called "inductive" or "nomological") science. Whereas nomological or inductive science addresses questions of the form, "How does nature normally operate or function?" historical science addresses questions of the form, "How did natural feature X arise?"[11]

This distinction has important implications for evaluating the scientific status of theories that invoke an antecedent cognitive act as a scientific explanation. I personally think that it suggests the legitimacy of such postulations if they also possess features such as wide explanatory power, internal consistency, and coherence. Let me explain.

When a research program concentrates on questions about how nature normally (i.e., unassisted by agency) operates, any reference to agency (whether divine or human) becomes inappropriate because it fails to address the question of interest. As I have argued elsewhere,[12] much nonhistorical scientific endeavor typically seeks to infer or explain nomological relations (i.e., scientific laws), whereas historical sciences typically seek to infer past causal events. To propose a divine act (construed as an event in space and time) where a nomological relation or law is required is to misunderstand the context and character of the relevant inquiry. Neither divine nor human action qualifies as a law. To offer either when a law is sought is clearly inappropriate. On this I believe, both theists (such as myself) and others (such as Professor Ruse) can agree.

It is not at all clear, however, that references to agency are similarly inappropriate when reconstructing a causal history—i.e., when attempting to answer questions about how a particular feature in the natural world (or the natural world itself) arose. In the first place, many fields of inquiry routinely invoke the action of agents to account for the origin of features or events within the natural world. Forensic science, history, and archaeology, for example, all sometimes postulate the past activity of human agents to account for the emergence of particular objects or events. Several such fields suggest that a clear precedent exists for inferring the past causal activity of intelligent agents as part of historical inquiry. (Imagine the absurdity of someone claiming that scientific method had been violated by the archaeologist who first inferred that French cave paintings had been produced by human beings rather than by natural forces such as wind and erosion.)

There is a second reason that postulating the past action of agency may be appropriate in the historical sciences. That has to do with the nature of historical explanations. Historical explanations require the postulation of antecedent causal events; they do not seek to infer laws (though they may use laws to make retrodictive inferences or to enhance the plausibility of a postulated causal history).[13] To offer past agency as part of an origins scenario or explanation is therefore (at least) logically appropriate, because the type of theoretical entity provided corresponds to the type required by historical explanations. Simply put, past agency is a causal event.[14] Agency, therefore, whether seen or unseen, may serve as a valid theoretical entity in a historical theory, even if it could not do so in a nomological or inductive one. Mental action may be a cause, even if it is certainly not a law.

I would like to press my case against Professor Ruse's prohibition against agency in science even further. I would like to argue that to exclude intelligent design *a priori* as a working hypothesis in, for example, historical biology is both gratuitous and anti-intellectual. Unlike Darwin, modern Darwinists can scarcely bring themselves to consider the possibility of intelligent design, let alone actually argue against it as he did. Professor Ruse, who to his credit has spent many hours directly confronting various creationist heresies, fails in this paper to mention intelligent design on his list of scientific possibilities. Yet it must be mentioned that this is precisely the theory that Darwin himself spent most of his time arguing against.

Indeed, it must be acknowledged that it is at least logically possible that a personal agent existed before the appearance of the first life on earth. It is therefore at least logically possible that such an agent (whether visible or invisible) designed or influenced the origin of life on earth. Moreover, as Bill Dembski will argue, we do live in the sort of world where knowledge of such an agent could in principle be accessible empirically. This suggests that it is logically and empirically possible that such an agent (whether divine or otherwise) designed or influenced the origin of life on earth. To insist that postulations of past agency are inherently unscientific in the historical sciences (where the express purpose of such inquiry is to determine what happened in the past) suggests we know that no personal agency could have existed prior to man. Not only is such an assumption intrinsically unprovable, it seems entirely gratuitous in the absence of some noncircular account as to why science should presuppose metaphysical naturalism.

Moreover, to exclude by assumption a logically possible answer to the question motivating historical science seems anti-intellectual and theoretically limiting, *especially since no equivalent prohibition exists on the possible nomological relationships that scientists may postulate in nonhistorical sciences.* The (historical) question that must be asked about biological origins is not "Which materialistic scenario will prove adequate?" but "How did life as we know it actually arise on earth?" Since one of the logically appropriate answers to this latter question is that "Life was designed by an intelligent agent that existed before the advent of humans," I believe it is anti-intellectual to exclude the "design hypothesis" without consideration of all the evidence, including the most current evidence, that might support it.

There is one final reason that *a priori* exclusions of design are anti-intellectual, indeed, even unscientific. Recent nonpositivistic accounts of scientific rationality suggest that scientific theory evaluation is an inherently comparative enterprise. Notions such as consilience[15] (which Professor Ruse mentions) and Peter Lipton's *Inference to the Best Explanation*[16] (IBE) imply the need to compare the explanatory power of competing hypotheses and/or theories. If this process is subverted by metaphysical gerrymandering, the rationality of scientific practice is vitiated. Theories that gain acceptance in artificially constrained competitions can claim to be neither "most probably true" nor "most empirically adequate." Instead such theories can be considered only "most probable or adequate among an artificially limited set of options." Moreover, where origins are concerned, only limited numbers of basic research programs are logically possible, as Professor Ruse mentions. (Either brute matter has the capability to arrange itself into higher levels of complexity or it does not, and if it does not, then either some external agency has assisted the arrangement of matter or matter has always possessed its present arrangement.)

The exclusion of one of the logically possible programs of origins-research by assumption, therefore, seriously diminishes the significance of any claim to theoretical superiority by advocates of a remaining program. Professor Ruse's prohibitions notwithstanding, an openness to empirical arguments for design is a necessary condition of a fully rational historical biology. In my opinion, a rational historical biology therefore must address not only the question, "Which materialistic evolutionary scenario provides the most adequate explanation of biological complexity?" but also the question, "Does a strictly materialistic evolutionary scenario, or one involving intelligent agency, or some other, best explain the origin of biological complexity, given all relevant evidence?" To insist otherwise is to insist that materialism holds a metaphysically privileged position. Since I see no reason to concede that assumption, I see no reason to concede Professor Ruse's conception of science.

The Fact of Evolution

For me, two things follow from the inadequacy of Professor Ruse's definition of science. First, because I reject Professor Ruse's view of science, I am unmoved by other similar philosophical arguments (especially from scientists) against the appropriateness of design theories in general. Indeed, almost all philosophical objections to the scientific status of intelligent design are predicated upon some untenable neo-positivist criterion of proper scientific practice. Many are predicated upon precisely the same "covering law" view of science that Professor Ruse has espoused.[17] Given recent work in philosophy of science by Laudan and others,[18] I doubt that Professor Ruse can offer a credible and metaphysically neutral demarcation criterion that succeeds in defining science narrowly enough to exclude the possibility of a scientific theory of design without also excluding evolutionary theories such as common descent.

Second, because I reject Professor Ruse's view of science, I am also unconvinced by his assurances that common descent is a fact, or as he once put it, "a fact, fact, FACT!"[19] I say this with no particular glee or malice, since I personally could quite easily accommodate common descent to my own belief that life owes its origin in some measure to intelligent design. I am simply unconvinced by the arguments for descent and by the philosophy of science that Professor Ruse and others invoke to make their case for it.

It might seem that Professor Ruse's philosophy of science and his arguments for common descent are unrelated. In fact, they are not. He acknowledged as much in his paper when he stated that "If you think in terms of unbroken law, then evolution makes the most sense." What if you don't think in terms of unbroken law—is common descent still a fact? Or rather, what if you reject the covering law model that leads Professor Ruse to speak of "unbroken law"—does common descent remain a fact then? Is it even still the best explanation? The fact is that common descent is not a fact, and Professor Ruse is letting his philosophical predilections about the nature of acceptable science drive his conclusions about biological history.

Strictly speaking, common descent is an abductive or historical inference,[20] as Professor Ruse himself acknowledges when he speaks more accurately of "inferring historical phylogenies." As defined by C. S. Peirce, abductive inferences attempt to establish past causes by viewing present effects.[21] Hence it is more accurate to refer to common descent as a theory about facts, i.e., a theory about what in fact happened in the past. Unfortunately, such theories, and the inferences used to construct them, can be notoriously underdetermined.[22] As Elliot Sober points out, many possible pasts often correspond to any given present state. Establishing the past with certainty, or even beyond reasonable doubt, can therefore, be very difficult—especially when the past in question occurred billions of years ago. In my opinion, none of Darwin's five main arguments for descent—neither fossil progression, biogeographical distribution, homology, embryological similarity, nor the existence of rudimentary organs[23]—establish common descent beyond reasonable doubt, though I admit that some of those arguments do strongly suggest the common ancestry of many disparate organisms within limited groups.

I also admit that the theory of common descent produces an admirable consilience. But that is just the point. *Theories* have the property of consilience; *facts* do not. In any case, consilience is a comparative notion, and to my mind the question of whether or not a monophyletic view of biological history can achieve a greater consilience than a polyphyletic view has not yet been settled. Indeed, even supposedly invincible arguments from molecular homologies depend for their efficacy upon *a priori* certainty that similarity cannot be the product of common principles of design. Such certainty in my experience often seems to have been acquired on the basis of rather naive dismissals of the metaphysics of others. It also seems to me to have been acquired without adequate reflection upon the implications of the molecular biological revolution which is now again suggesting to many of us the possibility of design.

NOTES

[1]Thomson (1982), pp. 529-531.

[2]In addition to his conference paper, see Ruse (1982b), pp. 72-78.

[3]Ruse (1986), pp. 68-73, especially footnote #9, p. 73. Ruse (1988), p. 301. Hempel, (1942), pp. 35-48. Hempel (1962), pp. 9-33.

[4]See, for example, Lipton (1991), pp. 43-46. Meyer (1990), pp. 39-76. Graham (1983), pp. 16-41. Scriven (1966), pp. 238-264. Mandelbaum (1961), pp. 229-242. Scriven (1959a), pp. 477-482. Scriven (1959b), pp. 448-451.

[5]Newton (1958), p. 302.

[6]Meyer (1990), pp. 47-75.

[7]*Ibid.*, pp. 51-56. Scriven (1975), p. 14. Lipton (1991), pp. 47-81.

[8]Scriven (1959b), pp. 446-463.

[9]Alston (1971), pp. 17-24.

[10]*Ibid.*, p. 17.

[11]For a thorough exposition of this, see Meyer (1990), pp. 1-136.

[12]*Ibid.*

[13]Indeed, none of the above denies that laws or process theories may play necessary roles in support of causal explanation, as even opponents of the covering-law model (such as Scriven) admit. Scriven notes that laws (or other types of general process theories) may play an important role in justifying the causal status of an explanatory antecedent and may provide the means of inferring plausible causal antecedents from observed consequents. Scriven (1959b), pp. 448-449; (1959a), p. 480;

(1966), pp. 249-250. See also Meyer (1990), pp. 18-24, 36-72, 84-92.

[14]For a more complete discussion of the prevailing neo-positivistic confusion of laws and causes, and the subsidiary role that nomological understanding does play in historical science, again, see Meyer (1990), pp. 36-76.

[15]Thagard (1978), p. 79. Whewell (1840), vol. 2:242. Gould (1986), p. 65. Laudan (1971), pp. 371-378.

[16]Lipton (1991), pp. 82ff.

[17]Ruse (1986), p. 73, especially footnote #9.

[18]See also Gillespie (1979), pp. 1-18, 41-66, 146-156. Saunders and Ho (1982), pp. 179-196. Quinn (1984), pp. 32-53. Laudan (1988), pp. 337-350. Meyer (1990), pp. 111-136. Lipton (1991). The untenable nature of Ruse's position is manifest in his own admission that modern evolutionary theory does not meet the demarcation standards that he promulgates elsewhere as normative for his opponents. See, for example, his discussion of population genetics in *Darwinism Defended* [Ruse (1982a), p. 86] where he acknowledges that "it is probably a mistake to think of modern evolutionists as seeking universal laws, at work in every situation."

[19]Ruse (1982a), p. 58.

[20]Meyer (1990), pp. 112-130. Gould (1986), pp. 60-69.

[21]Meyer (1990), pp. 24-34. Fann (1970), p. 33. Peirce (1931), vol. 2:375. Peirce (1956), pp. 150-156.

[22]Sober (1988), pp. 1-17.

[23]Ho (1965), pp. 8-20. Darwin (1859), pp. 331-434.

BIBLIOGRAPHY

Alston, W. P. (1971) "The place of the explanation of particular facts in science." *Philosophy of Science* 38:13-34.

Darwin, C. (1859) *The Origin of Species by Means of Natural Selection*. London. All quotes taken from version reprinted by Penguin in Harmondsworth, England, 1984.

Fann, K. T. (1970) *Peirce's Theory of Abduction*. The Hague: Martinus Nijhoff.

Gillespie, N. C. (1979) *Charles Darwin and the Problem of Creation*.

Chicago: University of Chicago Press.

Gould, S. J. (1986) "Evolution and the triumph of homology, or why history matters." *American Scientist* 74:60-69.

Graham, G. (1983) *Historical Explanation Reconsidered.* Aberdeen: Aberdeen University Press.

Hempel, C. (1942) "The function of general laws in history." *Journal of Philosophy* 39:35-48.

Hempel, C. (1962) "Explanation in science and in history." In: *Frontiers of Science and Philosophy*, R. Colodny (ed.), pp. 9-33. Pittsburgh.

Ho, Wing Meng (1965) *Methodological Issues in Evolutionary Theory.* D.Phil. thesis, Oxford University.

Laudan, L. (1971) "William Whewell on the consilience of inductions." *The Monist* 55:368-391.

Laudan, L. (1988) "The demise of the demarcation problem." In: *But Is It Science?* Ruse, M. (ed.), pp. 337-350. Buffalo: Prometheus Books.

Lipton, P. (1991) *Inference to the Best Explanation.* London: Routledge.

Mandelbaum, M. (1961) "Historical explanation: the problem of covering laws." *History and Theory* 1:229-242.

Martin, R. (1972) "Singular Causal Explanation." *Theory and Decision* 2:221-237.

Meyer, S. C. (1990) "Of clues and causes: a methodological interpretation of origin of life studies." Cambridge University Ph.D. thesis.

Newton, Isaac (1958) *Isaac Newton's Papers and Letters on Natural Philosophy.* I. Bernard Cohen (ed.), Cambridge: Harvard University Press.

Peirce, C. S. (1931) *Collected Papers*, vols. 1-6. C. Hartshorne and P. Weiss (eds.). Cambridge: Harvard University Press.

Peirce, C. S. (1956) "Abduction and Induction." In: J. Buchler (ed.) *The Philosophy of Peirce*, pp. 150-156. London: Routledge.

Ruse, M. (1973) *The Philosophy of Biology.* London.

Ruse, M. (1982a) *Darwinism Defended: A Guide to the Evolution Controversies.* London: Addison-Wesley.

Ruse, M. (1982b) "Creation science is not science." *Science,*

Technology and Human Values, vol. 7, no. 40, pp. 72-78.

Ruse, M. (1986) "Commentary: the academic as expert witness." *Science, Technology and Human Values*, vol. 11, no. 2, pp. 66-73.

Ruse, M. (1988) "Witness testimony sheet: McLean vs. Arkansas." In: *But Is It Science?* Ruse, M. (ed.), pp. 301-306. Buffalo: Prometheus Books.

Saunders, P. T., and Ho, M. W. (1982) "Is Neo-darwinism falsifiable?—and does it matter?" *Nature and System* 4:179-196.

Scriven, M. (1959a) "Explanation and prediction in evolutionary theory." *Science* 130:477-482.

Scriven, M. (1959b) "Truisms as the ground for historical explanations." In: *Theories of History*, P. Gardiner (ed.), pp. 443-475. Glencoe: The Free Press.

Scriven, M. (1966) "Causes, connections and conditions in history." In: *Philosophical Analysis and History*, W. Dray (ed.), pp. 238-264. New York.

Scriven, M. (1975) "Causation as explanation." *Nous* 9:3-15.

Sober, E. (1988) *Reconstructing the Past.* Cambridge: M.I.T. Press.

Thomson, K. S. (1982) "The meanings of evolution." *American Scientist* 70:529-531.

Whewell, W. (1840) *The Philosophy of the Inductive Sciences,* 2 vols. London.

II
Johnson-Ruse Debate

"Can Darwinism be Reconciled with Any Meaningful Form of Theistic Religion?"

Darwinism and Theism
Phillip E. Johnson

Theism and Darwinism:
Can You Serve Two Masters at the Same Time?
Michael Ruse

4
Darwinism and Theism
Phillip E. Johnson

SINCE THE PUBLICATION OF *Darwin on Trial*, friends have been sending me copies of a newsletter called *BASIS*, mainly because it often has something unfavorable to say about me. *BASIS* is published by an organization calling itself the San Francisco Bay Area Skeptics. As you can imagine, these Skeptics do not encourage people to be skeptical about doctrines of the rationalist faith like atheism, materialism, and Darwinian evolution. A recent issue of *BASIS* reported on a local meeting at which the featured speaker was a woman identified as "a religious person and science teacher at a Catholic school." This science teacher was assuring her audience that, despite the religious affiliation of her school, she taught evolution and not creationism in her science classes. A questioner from the audience then put her on the spot by asking, "Do you think that evolution is directed?" The newsletter reports that this question was followed by a "dramatic pause," after which the teacher replied with what it called a "battled 'No.'" The reporter for *BASIS* commented, "I would have expected a more rapid answer, but the battle between her curriculum and her beliefs had a few more moments of unrest left to settle."[1]

That conflict symbolizes for me the quandary of all those scientifically literate Christian intellectuals who struggle to reconcile Darwinism and theistic religion. Most of these people would probably call themselves theistic evolutionists. The name implies that they consider evolution to be a process initiated and guided by God, presumably in order to bring about the existence of human beings. My impression is that most theistic evolutionists in their hearts think of evolution as God's chosen means of creation, although in their heads they know that this concept is more a form of "soft creationism" than genuine evolutionism as Darwinist scientists use the term. The tension between head and heart leads to a characteristic vagueness when theistic evolutionists try to explain exactly what God had to do with evolution. From the hesitancy of that teacher's response to the crucial question, I suspect that she probably did not go out of her way at that Catholic school to call the attention of her students, and especially their parents, to the unanimity with which contemporary Darwinist authorities repudiate the idea that evolution is directed by any supernatural intelligence. A representative statement, typical of the official Darwinist attitude, is this one by George Gaylord Simpson:

> Although many details remain to be worked out, it is already evident that all the objective phenomena of the history of life can be explained by purely naturalistic or, in a proper sense of the sometimes abused word, materialistic factors. They are readily explicable on the basis of differential reproduction in populations (the main factor in the modern conception of natural selection) and of the mainly random interplay of the known processes of heredity. . . . Man is the result of a purposeless and natural process that did not have him in mind.[2]

The leading Darwinist authorities are frank about the incompatibility of their theory with any meaningful concept of theism when they are in friendly territory, but for strategic reasons they sometimes choose to blur the message. When social theorist Irving Kristol published a *New York Times* column in 1986 accusing Darwinists of manifesting a doctrinaire antitheism, for example, Stephen Jay Gould responded in *Discover* magazine with a masterpiece of misdirection.[3] Quoting nineteenth century preacher Henry Ward Beecher, Gould proclaimed that "Design by wholesale is grander than design by retail," neglecting to inform his audience that Darwinism repudiates design in either sense. To prove that Darwinism is not hostile to "religion," Gould cited the example of Theodosius Dobzhansky, whom he described as "the greatest evolutionist of our century, and a lifelong Russian Orthodox." As Gould knew very well, Dobzhansky's religion was evolutionary naturalism, which he spiritualized after the manner of Pierre Teilhard de Chardin. A eulogy published by Dobzhansky's pupil Francisco Ayala in 1977 described the content of Dobzhansky's religion like this:

> Dobzhansky was a religious man, although he apparently rejected fundamental beliefs of traditional religion, such as the existence of a personal God and of life beyond physical death. His religiosity was grounded on the conviction that there is meaning in the universe. He saw that meaning in the fact that evolution has produced the stupendous diversity of the living world and has progressed from primitive forms of life to mankind. Dobzhansky held that, in man, biological evolution has transcended itself into the realm of self-awareness and culture. He believed that somehow mankind would eventually

evolve into higher levels of harmony and creativity.[4]

Evolution is thoroughly compatible with religion—when the object of worship is evolution.

I don't mean to pick on Gould, because in being evasive about the implications of Darwinism for religion he was merely following the lead of the prestigious National Academy of Sciences. In an official 1984 statement the Academy's president assured the public that it is "false . . . to think that the theory of evolution represents an irreconcilable conflict between religion and science." Dr. Frank Press explained:

> A great many religious leaders accept evolution on scientific grounds without relinquishing their belief in religious principles. As stated in a resolution by the Council of the National Academy of Sciences in 1981, however, "Religion and science are separate and mutually exclusive realms of human thought whose presentation in the same context leads to misunderstanding of both scientific theory and religious belief."[5]

That statement could have been drafted by one of those White House or Congressional "spin doctors" whose assignment is to mislead the public without telling an outright lie. Dr. Press did not say whether the religious leaders in question were simply overlooking a logical contradiction, or whether the "religious principles" they managed not to relinquish included a creating God who takes an active role in designing or constructing living organisms. He also did not say what the compulsory separation of science and religion implies for those scientists who continually make purportedly scientific statements about the purposelessness of evolution or the absence of a supernatural creator from the history of the cosmos. No wonder the candid scientific materialist William Provine described the National Academy's position as politically understandable but intellectually dishonest.[6]

The present discussion is over whether belief in Darwinism is compatible with a meaningful theism. When most people ask that question, they take the Darwinism for granted and ask whether the theism has to be discarded. I think it is more illuminating to approach the question from the other side. Is there any reason that a person who believes in a real, personal God should believe Darwinist claims that biological creation occurred through a fully naturalistic evolutionary process? The answer is clearly *No*. The sufficiency of any process of chemical evolution to produce life has not been demonstrated, nor has the ability of natural selection to produce new body plans, complex organs, or anything else except variation within types that already exist. Papers presented at this symposium explain why Darwinian innovation of this sort is exceedingly unlikely. The fossil record does not evidence any continuous process of gradual change, which is why paleontologists are continually tempted to flirt with the heresy that biological transformations occurred in sudden jumps. If chemical and biological evolution is the only possible source of living organisms, then the shortage of evidence is of little importance; the only question is how naturalistic evolution occurred, not whether it did. If God exists, then naturalistic evolution is not the only alternative, and there is no reason for a theist to believe that God employed it beyond the relatively trivial level where variation has been demonstrated.

From a theistic perspective, Darwinism as a general theory is not empirical at all. It is a child of naturalistic or positivistic philosophy, which defines science as the attempt to explain the world without allowing any role to theological or providential activity. Positivism in this sense requires science to have at least a vague theory about everything really important. To produce the required theory, scientists are allowed, if necessary, to make simplifying assumptions or even to overlook difficult aspects of the problem. Even a particularly frustrating problem, such as the origin of life on earth, is considered to be solved in principle once scientists think they have some plausible general notion about how the thing might have happened. The spirit of positivistic science is illustrated by James Trefil's summary of the evolution of life in his recent book, *1000 Things Everyone Should Know About Science*:

> Evolution of life on earth proceeded in two stages: chemical and biological. Life on earth must have developed from inorganic materials—what else was there for it to come from? The first stage in the development of life, therefore, was the production of a reproducing cell from materials at hand on the early earth. This process is called chemical evolution. . . . Once a living, reproducing system was present, the process of natural selection took over to produce the wide variety of life that exists today.[7]

That sort of reasoning seems unimpeachable to metaphysical naturalists; fully naturalistic chemical and biological evolution happened because nothing else could have happened. A theist, on the other hand, has no reason to accept the plausibility of either chemical evolution or creative natural selection in the absence of a convincing empirical demonstration.

Because Darwinism has its roots in metaphysical naturalism, it is not consistent to accept Darwinism and then to give it a theistic interpretation. Theistic evolutionists are continually confused on this point because they think that Darwinism is an empirical doctrine—i.e., that it rests fundamentally on observation. If that were the case, it is hard to see how any observations of evolution or natural selection in action could rule out the possibility that Darwinian evolution is God's way of creating. Nothing about the observed variations in the beaks of finches in the Galápagos Islands, or in the increased survival rate of dark melanic moths during periods when the background trees were darkened by industrial smoke, discredits a theistic interpretation of evolution. If one assumes that confidence in the ability of Darwinian selection to create entirely new kinds of animals is based on observations like those, then obviously atheism or metaphysical naturalism is not a necessary implication of Darwinism. This mistaken premise leads theistic evolutionists to the conclusion that they can accept George Gaylord Simpson's "scientific" statement—i.e., that mutation and selection did the work of creation—and reject his "philosophical" conclusion that the universe is purposeless.

The flaw in that logic is that the purportedly scientific statement was inferred from the philosophical conclusion rather than the other way around. The empirical evidence in itself is inadequate to prove the necessary creative power of natural selection without a decisive boost from the philosophical assumption that only unintelligent and purposeless processes operated in nature before the evolution of intelligence. Darwinists know that natural selection created the animal groups that sprang suddenly to life in the Cambrian rocks (to pick a single example) not because observation supports this conclusion but because naturalistic philosophy permits no alternative. What else was available to do the job? Certainly not God—because the whole point of positivistic science is to explain the history of life without giving God a place in it.

In short, the reason that Darwinism and theism are incompatible is not that God could not have used evolution by natural selection to create. Darwinian evolution might seem unbiblical to some, or an unlikely method for God to use, but it is always possible that God might do something that confounds our expectations. The contradiction between Darwinism and theism is at a deeper level. To know that Darwinism is true (as a general explanation for the history of life), one has to know that no alternative to naturalistic evolution is possible. To know *that* is to know that God does not exist, or at least that God cannot create. To infer that Darwinism is true because there is no creator God, and then to interpret Darwinism as God's method of creating, is to engage in self-contradiction.

I have two concluding points. First, the contradiction between Darwinism and theism is not necessarily evident to people who have only a superficial acquaintance with Darwinism. That explains why 40 percent of the American public believes in a God-guided evolution and thinks, no doubt, that this position satisfactorily reconciles science and religion. The contradiction sinks in when a person assimilates Darwinist ways of thinking and sees how antithetical they are to theism. That is why Darwin in his own time and his successors today have generally felt that theistic evolutionists were missing the point.[8] Theistic evolutionists protest (correctly) that a creative role for natural selection does not rule out the possibility of God, but they fail to understand that the entire outlook of positivistic science is profoundly incompatible with the existence of a supernatural creator who takes an active role in the natural world.

My second concluding point is that it is risky for Darwinists to be candid about the implications of their theory for theistic religion. I don't mean simply that the anti-theistic bluster put about by people like William Provine and Carl Sagan arouses opposition, although that is an important consideration. I am thinking of an intellectual problem. The all-purpose defense that Darwinists invoke when their theory is under attack is to invoke what I called in my earlier address "Dobzhansky's rules," the rules of positivistic science. That is, they say that "science" is defined as the search for naturalistic explanations for all phenomena and that any other activity is "not science." This position is sustainable only on the assumption that "science" is just one knowledge game among many, and theists suffer no great loss if they have to go and play in another game called "religion." The problem is that the games do not have equivalent status. The science game has government support and control of the public educational establishment. Everybody's children, theists and non-theists alike, are to be taught that "evolution is a fact." This implies that everything contrary to "evolution," specifically the existence of a God who takes a role in creation, is false. If "evolution" has strong anti-theistic implications, the theists in the political community are entitled to ask whether what Darwinists promulgate as "evolution" is really true. The answer, "That's the way we think in science," is not an adequate response.

In the famous Arkansas creationism trial, the Darwinist expert witnesses were able to lead the gullible Judge William Overton by the nose and persuade him that theists have no legitimate intellectual objection to the Darwinist world view. As authority for the proposition that belief in a divine creator and acceptance of the scientific theory of evolution (i.e., Darwinism) are compatible, Judge Overton cited none other than Francisco Ayala, author of the previously quoted eulogy of Theodosius Dobzhansky.[9] The next time this sort of issue comes around, I predict that the Darwinists will have to deal with a more sophisticated judicial audience.

NOTES

[1]"Stockton: Report from the Front Lines of Public School Science Education," Commentary, *BASIS* (December 1991), p. 3.

[2]George Gaylord Simpson, *The Meaning of Evolution* (rev. ed., 1967), pp. 344-345.

[3]Stephen Jay Gould, "Darwinism Defined: The Difference Between Fact and Theory," *Discover* (January 1987), pp. 64-70.

[4]Francisco Ayala, "Nothing in biology makes sense except in the light of evolution," *Journal of Heredity*, vol. 68 (January-February 1977), pp. 3, 9.

[5]The official position paper on creationism of the National Academy of Sciences was published in 1984, with beautiful illustrations, under the title *Science and Creationism: A View from the National Academy of Sciences*. The paper was prepared under the direction of a distinguished committee of seven scientists and four lawyers. The quotation in my text is from the introduction to the paper, signed by the Academy's President, Dr. Frank Press.

[6]William Provine, "Evolution and the Foundation of Ethics," *MBL Science* (publication of the Marine Biological Laboratory at Woods Hole, Massachusetts), vol. 3, no. 1, pp. 25-29.

[7]Stephen Compton, "From Electricity to Chaos," *San Francisco Chronicle*, Book Review Section (February 23, 1991), p. 7.

[8]See Neal C. Gillespie's account of Darwin's frustration with the theistic evolutionists of his own time in *Charles Darwin and the Problem of Creation* (University of Chicago Press, 1979), especially chapter 5.

[9]Footnote 23 in Judge Overton's opinion states that "The idea that belief in a creator and acceptance of the scientific theory of evolution are mutually exclusive is a false premise and offensive to the religious

views of many . . . . Dr. Francisco Ayala, a geneticist of renown and a former Catholic priest who has the equivalent of a Ph.D. in theology, pointed out that many working scientists who subscribe to the theory of evolution are devoutly religious." McLean v. Arkansas, 1529 F. Supp. 1255 (W.D. Ark. 1982). Reprinted in the collection *But Is It Science?* (Ruse, ed., 1988), p. 330. It does not seem to have occurred to Judge Overton to wonder why Dr. Ayala is a *former* Catholic priest.

5

Theism and Darwinism:
Can You Serve Two Masters at the Same Time?
Michael Ruse

CAN A THEIST BE a Darwinian? Can a Darwinian be a theist? People always complain that philosophers are obsessed with words, and there is some truth in that. Sometimes, however, you can avoid a great deal of cross-talk by looking carefully, at the beginning of the discussion, at the terms you are going to use. So, without further apology, to answer my questions, let us start by teasing out some meanings to the terms *theism* and *Darwinian*. I want to emphasize that I look upon this discussion as a prolegomenon to decision-making; at no time shall I be saying what is right or wrong, best or worst. I am simply trying to lay out the options. I function as a bureaucrat, not as an advocate.[1]

What Do You Mean by "Theist"?

I take it that a "theist" is a person who believes in a god, a god who is prepared to intervene in his (her/its) creation. This is compared to a "deist," a person who believes in a god that is not prepared to intervene in its creation, an "Unmoved Mover." Both of these are compared to an "agnostic," who professes ignorance about the deity, and an "atheist," who does not believe in the existence of a god at all.

Traditionally, theists have been thought of as belonging to one of the great religions of the Mediterranean: Judaism, Christianity, and Islam. Other religions are "pagan"—although I see no reason in principle why they should not qualify as theistic or deistic. To keep my discussion within bounds, I shall confine my discussion to Christian theism. If you fault me for ethnocentrism, I shall have no reply.

In a sense, there are almost as many notions of Christian theism (Christianity, for short, from now on) as there are Christians. Cutting across all divisions, including in a fashion the division between Catholics and Protestants (not to mention the Orthodox), I shall distinguish three levels or grades. I shall call them *conservative* Christianity, *moderate* Christianity, and *liberal* Christianity. I think that I can use these terms without undue distortion, but please do not fault me if, say, the present Pope, whom you would judge a conservative, comes out as a moderate on my schema. I do not intend the grades to be sharp at the boundaries. In real life, people might be conservative in one sense, and moderate in another; or they might fall on a dividing line.

For me, a conservative Christian is one who takes the truth of the Bible, and/or the teaching of the church (often, but not necessarily, the Catholic church) fairly literally. I am not saying that the conservative necessarily has to take every last word of the Bible as the unalterable, face-value truth—since Augustine, Catholicism has had a tradition of interpretation—but I assume that unbending literalists ("fundamentalists") are all conservatives in my sense.

For me, therefore, a conservative will believe in a real garden of Eden, a real Adam and Eve, and a real Fall. A conservative will believe in a real flood, although I can imagine that he or she might not really care if the flood failed to reach as far as Texas. A conservative will believe that Jesus Christ was the son of God, that he performed the biblical miracles (and that they were genuine miracles), that he died for our sins, that his body started to stink, and that then he rose from the dead, joining God in heaven—where some of us might hope to go to share eternal bliss. I really do not know where today's conservative stands on hell (burning flames or nonbeing), but he/she believes that that is the punishment for the sinner.

My moderate Christian believes much that the conservative believes—for instance, that actual people sinned, that Jesus was genuinely the son of God, that he performed miracles, that he rose from the literal dead, and that there is salvation for the repentant sinner. I doubt, however, that my moderate is going to spend funds and time trying to find the true home of Eden, or the remains of Noah's ark. My moderate likewise might wonder if one has to follow slavishly every dictate of St. Paul, sensing sometimes that the apostle told more of his own psyche than of God's wishes.

My liberals, perhaps, technically ought to be thought of as deists and not as theists, but for sociological reasons, if for no other, they can be included here. The liberal is one who interprets the Bible and church teaching in modern terms. Most of the stories of the Old Testament are taken to be allegorical, the miracles of Jesus are given natural explanations (if they are believed at all), and much effort is put into showing that the resurrection does not necessarily imply bodily resurrection. Original sin is thought to be something inherent in us all, and not necessarily the consequence of our first parents' failing.

These then are my three types of Christian. I have emphasized that I am being nonjudgmental. What I would stress is that it is possible to find, in all three levels, people who are genuinely committed to their faith. The conservative might think the liberal no true Christian. I can testify that there are extreme liberals who are as devoted to their savior as any fundamentalist, and who find their faith a great deal more difficult and demanding than do most. Conversely, I know conservatives who have made very real sacrifices for what they believe to be the truth.

What Do You Mean by "Darwinian"?

As a fairly enthusiastic Darwinian myself, I can attest to the fact that "Darwinism," no less than "Christianity," is a notion with many meanings. Again, I will propose three grades. In a sense, these correspond to my three grades of Christian. I feel a bit diffident, however, about referring to an ardent Darwinian as a "conservative Darwinian." That is a misnomer, if not an oxymoron. Hence, I shall speak of the *ultra*-Darwinian, the *moderate* Darwinian, and the *restrained* Darwinian. These are not necessarily the most elegant terms, but they will serve.

All three Darwinians are evolutionists, believing that organisms, including ourselves, came by a process of development from a few simple forms. The ultra-Darwinian thinks that the sole cause was Charles Darwin's mechanism of natural selection working on random (not uncaused) variations. This factor suffices to explain all. There are no other causes at work, nor are other causes needed. This means that all organic features are to be considered adaptive, even though we may not at present know precisely the nature of the function of these adaptations.

The classic problem case is that of male nipples. What function could these possibly serve? The ultra-Darwinian thinks that they have to have some end, like sexual attractiveness. An explanation in terms of being a byproduct of other features, or some such thing, will not do. I do not know how many people are ultra-Darwinians of this extreme ilk today, but they have certainly existed in the past. Alfred Russel Wallace, the co-discoverer of natural selection, was one before his conversion to spiritualism. The turn of the century biometrician, Raphael Weldon, was another.

The moderate Darwinian thinks that natural selection is the most important mechanism of evolutionary change. But he or she is unwilling to give selection complete and exclusive causal authority over evolution. The moderate thinks that there might well be other causes of change which, in their way, could be very important. Included here are genetic drift, correlation of parts, and perhaps even "hopeful monsters." No one today believes in Lamarckism, in the old-fashioned sense of the direct inheritance of acquired characters. Some today think that non-Darwinian factors might be very important at the molecular level.

The restrained Darwinian thinks that selection is certainly at work and may have important effects. However, he/she does not think it the most important cause of change. We must look for other factors of change to explain the overall pattern. In the past, someone like the American paleontologist Henry Fairfield Osborn would have come under this heading. Today one might include the Harvard paleontologist Stephen Jay Gould in this category, although I myself think he is more properly labeled a moderate. (As with Christianity, I do not intend to imply that the categories of Darwinism are sharp and exclusive; some people will fall on the boundaries.)

Can a Christian Be a Darwinian?

Now that we have our terms spelled out, we can set about answering our question. The answer obviously is that it all depends on what you mean by "Christian" and what you mean by "Darwinian." So let us start running our different categories past each other.

Start with the conservative Christian. Where would he or she stand with respect to Darwinism of any variety? My feeling is that there would not be much sympathy for Darwinism at all, ultra, moderate, or restrained. If this conservative is an outright biblical literalist, I do not see how he/she could be an evolutionist at all; and, more important, I do not see that he/she would want to be an evolutionist anyway. His or her basic belief would be in a miraculous creation of life and of frequent divine interventions thereafter. The spirit of such an outlook is against a natural account of origins.

Would it be possible, nevertheless, for the conservative to be an evolutionist, supposing that one were prepared to allow a minimum amount of interpretation? Or, supposing that one really did not think that the Bible necessarily tells us about everything, could one accept some measure of development? I do not see why that would be impossible. I doubt that such a person would be much of a Darwinian; probably he or she would want some sort of directed evolutionism. Or, one might want to restrict change to that occurring within major types (within the reptiles, for instance). Certainly, the presupposition is that there are many significant miracles, which break with the laws of nature, whether those laws be evolutionary or not. But, within these strong bounds—accepting evolution as secondary, as it were—one could allow limited development.

Before you dismiss my suggestion as ridiculous, let me suggest that there have in fact been people who fit this category. Remember that fundamentalism is a very restricted version of Christianity, is essentially an American production, and is not that ancient. It is a child of the nineteenth century. But in that nineteenth century, one also had people like John Henry Newman, a Catholic convert and very conservative in much of his thinking. As a Catholic, Newman's first allegiance was to the church and not to the literal truth of the Bible. He is in fact on record as saying that if evolution be true, then so be it. Fundamentally, Newman was not interested in science; it neither helped nor hindered his religion. Hence, his attitude was that one should not pick a quarrel unnecessarily. He knew that his redeemer liveth, whatever the truth of evolution.

I come next to what I have called the moderate Christian. I think you might get some surprising answers here—at least surprising until you think about them. Clearly, the moderate Christian cannot be an ultra-Darwinian in the sense of allowing nothing but unbroken law at all times. The moderate believes in many of the biblical miracles, including the greatest of all, the resurrection of Jesus and the washing away of our sins. I suspect also that the moderate might have trouble, or certainly feel the need to think hard, concerning some other claims of the ultra-Darwinian (perhaps of the other kinds of Darwinian also).

I am thinking here particularly about the story of Adam and Eve and the Fall. One might not believe in a literal garden of Eden, but presumably one will believe that there was a first pair of humans and that they sinned. It is possible on Darwinian theory to think that you might get down to a bottleneck of just one pair—even just one fertilized female—and so presumably one could reconcile the Genesis story in that way. But I am not sure one has the right to think that this must have happened, in order to save one's science. Obviously one might try other options, for instance, assuming that God gave all extant humans immortal souls at one instant, and that then they sinned collectively or that the sin of one pair was transferred to all, or some such thing. The point is that one has got to think of something, and this might require a rethinking of one's theology—as long as one wants to stay with the science, that is.

On the other side, however, let me point out that the ultra-Darwinian argues that there are designlike effects throughout the living world. It is true that these come about through a struggle for existence, but the problem of evil is no stranger to the Christian. What is welcome to the Christian (one moderate enough to be an evolutionist of a kind) is that his/her natural theology is thus confirmed by the Darwinian, by the ultra-Darwinian especially. Hence, what I am suggesting is that even though the moderate Christian can hardly accept the full program of the ultra-Darwinian, in respects he/she is going to be drawn much more toward the ultra end than the restrained end or even the moderate middle of the Darwinian spectrum.

Again, I would point out that before you dismiss this as so much hypothetical theorizing, there have in fact been people who think this way—embracing a fairly strong moderate-to-conservative Christianity and yet drawn by natural theology to an ultra-Darwinian stance. The great evolutionist Sir Ronald Fisher was one. There were also those, especially conservative Presbyterians in the nineteenth century, who were drawn to ultra-Darwinism because the struggle leading to selection confirmed what they had always believed about God's separating the sheep and the goats, and his choosing only the former.

I come finally to the liberal Christian. As I have said, in some respects I see this person as being close to deism rather than theism. But however you categorize such a person, the fact is that he or she will positively welcome the advances of science, seeing in every new discovery fresh evidence of God's power at work and the triumph of his great gift to us, our ability to reason and understand. Evolution will be taken as one of the glories of science and as a testament to his greatness.

Whether such a Christian as this will be an ultra, moderate, or restrained Darwinian seems to me to be an open question, and I suspect that such a believer would incline to think such a question a little irrelevant. The matter at issue is God's power, as revealed through his law—and for this, any kind of naturalistic evolutionism is both necessary and sufficient. Indeed, if I were to hazard a guess, it would be that in respects the liberal Christian would feel less drawn to ultra-Darwinism than the moderate Christian, paradoxical though this suggestion may seem, simply because traditional natural theology, especially teleology, would have less of a hold on the liberal than on the moderate. I have in mind here someone like the Anglican priest Arthur Peacocke or the Lutheran pastor Philip Hefner. But, whatever the option taken, such a Christian would see Darwinism as supporting his/her faith, not threatening it.

A Final Word

There is no single answer to the question I posed at the beginning of this discussion. It all depends on what you mean by your terms, and what you mean can lead to diametrically opposing conclusions. Throughout, as promised, I have tried to be nonjudgmental. It is enough here to analyze the options. I trust that the worth of what I have done needs no justification above its execution.

But, as I conclude, let me say one final word. I speak now especially to those who hold strong opinions. Do not, I beg of you, assume without argument that you and your group, alone, have an exclusive lien on the truth or on the genuine religious spirit. You may be right, and you may be more holy than most; but remember that there are many people in different times and places—very different times and places, if you include non-Christians—who do not see things as you do. I say this, irrespective of whether you are a conservative, moderate, or liberal Christian, or not a Christian at all.

Above all, do not think people insincere if they do not solve the science/religion problem in the way that you do. Before you assume that your way of religious thought must be the proper and superior way, remember that it was not so very long ago that Michigan thought that it alone had the proper and superior way of making automobiles. I would not want you to end as the theological equivalent of General Motors.

NOTES

[1]It hardly seems necessary to load down so elementary a discussion as this with a massive number of notes and references. Two books that I have found very helpful are Ernan McMullin (ed.), *The Creation-Evolution Controversy* (Notre Dame: University of Notre Dame Press, 1986) and Arthur Peacocke, *God and the New Biology* (London: Dent, 1986). McMullin is a Catholic priest and Peacocke an ordained minister in the Church of England (Episcopalian). The closest I have come to talking about these matters is in my edited volume, *But Is It Science? The Philosophical Question in the Creation/Evolution Controversy*

(Buffalo: Prometheus Books, 1988). In the final essay of my *The Darwinian Paradigm* (London: Routledge, 1989), I raise some problems for the Christian about the question of the foundations of morality from a Darwinian perspective. My Christian friends all tell me that my worries are unfounded.

III
General Discussion, Responses, and Replies

6

Experimental Support for Regarding Functional Classes of Proteins to Be Highly Isolated from Each Other
Michael J. Behe

IN WRITING ON THE TOPIC of naturalism and evolution the problem arises of what to call the contending camps. The difficulty comes from the fact that although the term *evolutionist* is often used to refer to persons who demand the unrelenting application of physical laws to all phenomena in the universe, many other persons who are opposed to this view are perfectly willing to concede that a limited number of phenomena can be explained by Darwinistic principles. Similarly, although a term like *creationist* brings to mind champions of a young-earth theory, it is often applied to persons who do not defend that thesis but do contend that natural laws have at some point been superseded by a supernatural agency.

Since the focus of this symposium is the sufficiency of natural law, and in order to avoid the confusing terminology discussed above, in this essay I will use the term *believer* for those who believe in the universal application of natural law and the term *skeptic* for those who doubt it. This has the advantage of using terms for each side that the opposite side generally regards positively. Perhaps this will go a little way toward promoting the good will that this conference strives for.

Introduction
Several years ago the fossilized remains of an extinct species of whale were unearthed in the Zeuglodon valley of Egypt. The particular aspect of the fossil which excited archaeologists and science writers was the fact that the whale apparently had functional legs and feet. From the condition of the fossilized leg bones it could be discerned by trained eyes that the legs were well muscled and thus must have been actively used during the life of the whale. A *Washington Post* story describing the discovery included a drawing of both a modern whale and an ancient whale, showing the differences in their shapes but the similarities in their lengths. Also included in the illustration, down in the lower

righthand corner, was a drawing of an animal that looked for all the world like a scruffy dog. Underneath the dog was the caption "Mesonychid, the ancestor of the whales." In the story it was explained that

> Most researchers agree the earliest whales descended from a line of large carnivorous beasts the size of wolves and bears. These furry land mammals, known as mesonychids, ran around on four legs. But for unknown reasons, some mesonychids evolved into forms that returned to the sea, from which all life originally arose. The legs found on primitive whales are remnants from their time on land (July 13, 1990).

Even allowing for the enthusiasms of the popular press, the story reflects the way in which a theory, here evolution, is allowed to supply "facts" which the evidence in no way justifies. I discussed this article with my students in a course I teach for freshmen, entitled "Popular Arguments on Evolution." The course is intended to develop critical reasoning skills, using popular books that have opposing viewpoints on evolution as the vehicle. This past semester we read, side by side, Richard Dawkins's *The Blind Watchmaker* and Michael Denton's *Evolution: A Theory in Crisis*. This forced the students to argue over the meaning of observations, without the automatic social support that usually goes to proponents of evolution in academic settings. The students themselves, after reading the *Post* article, pointed out that there is no reason to suppose that the ancient whale appeared on earth before the modern whale, since modern whales have vestigial legs that could have developed into the functional legs of the Zeuglodon whale. For the same reason, the students noted, the discovery does not represent the development of a new trait or even the loss of an old one. Finally, most glaringly obvious, if random evolution is true, there must have been a large number of transitional forms between the *Mesonychid* and the ancient whale. Where are they? It seems like quite a coincidence that of all the intermediate species that must have existed between the *Mesonychid* and whale, only species that are very similar to the end species have been found. The students concluded that the fossil whale, although a fascinating discovery for natural history, was no evidence for the *Post*'s evolutionary scenario.

I have started my contribution to this symposium with a discussion of the Zeuglodon whale because it is a paradigmatic example of evolutionary argumentation: a small change in a preexisting structure is used to argue to massive changes involving completely new structures or functions. It is like arguing that because a man can jump over a fissure five-feet wide, then given enough time he could jump over the Grand Canyon. Now, a believer in the unabating rule of natural law would argue that the man could jump over the Grand Canyon if there were ledges and buttes for him to use as steppingstones. The skeptic would ask to be shown the steppingstones.

This essay will examine how the search is going for steppingstones in one area of biochemistry, that of protein structure. We will see that, without a prior commitment to naturalism, there is little reason to suppose that steppingstones exist in the canyon separating functional classes of proteins.

Protein Structure

I ask for the patience of those who already have a working knowledge of protein structure, but in order to make sure that everyone reading this essay has the necessary background I will spend a little time discussing some fundamentals.

Although most people think of proteins as something we eat—one of the major food groups—when they reside in the body of an uneaten animal or plant, proteins serve a different purpose. Proteins are the machinery of living tissue that builds the structures and carries out the chemical reactions necessary for life. For example, the conversion of foodstuffs to biologically usable forms of energy is carried out, step by step, by part of a group of proteins called enzymes. Skin is made in large measure of a protein called collagen. When light impinges on your retina it interacts first with a protein called rhodopsin.

As can be seen even by this limited number of examples, proteins carry out amazingly diverse functions. In general, however, a given protein can perform only one or a few functions: rhodopsin cannot form skin, and collagen cannot interact usefully with light. Therefore a typical cell contains thousands and thousands of different types of proteins to perform the many tasks necessary for life, much like a carpenter's workshop might contain many different kinds of tools for various carpentry tasks.

What do these versatile tools look like? The basic structure of proteins is quite simple: they are formed by hooking together in a chain discrete subunits called amino acids. Now, although the protein chain can consist of anywhere from about fifty to about one thousand amino acid links, each position can contain only one of twenty different amino acids. In this they are much like words: words can come in various lengths but they are made up from a discrete set of twenty-six letters. As a matter of fact, biochemists often refer to each amino acid by a single letter abbreviation: *G* for glycine, *S* for serine, *H* for histidine, and so forth. Each different kind of amino acid has a different shape and different chemical properties; for example, *W* is large but *A* is small, *R* carries a positive charge but *E* carries a negative charge, *S* prefers to be dissolved in water but *I* prefers oil, etc. A protein in a cell does not float around like a floppy chain; rather, it folds up into a precise structure that can be quite different for different types of proteins. This is done automatically through interactions such as a positively charged amino acid trying to get near a negatively charged one, oil-preferring amino acids trying to huddle together to exclude water, large amino acids being excluded from small spaces, etc. When all is said and done, two different amino acid sequences, two different proteins, can be folded to structures as specific and different from each other as a three-eighths inch wrench and a jigsaw. Like the household tools, if the shape of the proteins is significantly warped, they fail to do their jobs.

Proteins and Language

Because amino acid residues are often abbreviated by letters, because there is a similar number of letters and amino acids (twenty-six vs. twenty, respectively), and because a small protein consists of about one hundred amino acids, many commentators have likened a functional protein (i.e., one that has the correct shape to be able to do a particular job) to a functional sentence (i.e., one that obeys the rules of English grammar) of about one hundred letters. My students in "Popular Arguments on Evolution" found it interesting that both believers and skeptics used this kind of analogy in their writings, but that their reasonings brought them to opposite conclusions. The skeptic typically argues that a monkey banging away at a typewriter (monkeys and typewriters are very popular) would be unlikely to produce an intelligible, grammatically correct sentence like "Drop the anchor in one hour" in a reasonable length of time. Near misses don't count for the skeptic since the change of even one letter would break a spelling or grammar rule, or change the sense of the sentence. Needless to say, the hour would most likely pass, and the anchor remain undropped, before the monkey produced the correct sentence.

Believers in the universal application of physical law take a different approach with their monkey and typewriter. Their argument generally goes something like this. Suppose in his first try the monkey typed "bsqm dshcbbbk,RR .nsurlei aknex." Admittedly this is poor grammar, but it's the only sentence we've got. Since living systems reproduce, and since there is Darwinian competition, the bad sentence will be reproduced until a better one comes along. Now suppose in his second try the monkey typed a *p* in the fourth position and a *u* in the penultimate position. Well, since these are closer to the target sentence we will throw out the original sentence and keep "bsqp dshcbbbk,RR .nsurlei aknux." After a few more rounds perhaps the monkey has gotten a few more letters correct, say, a *d* in the first position and a *ch* in the thirteenth and fourteenth positions. Now we have "dsqp dshcbbbchRR .nsurlei aknux." Since this has more matches with the target sentence we'll keep it and throw out the last sentence. After perhaps fifty rounds we get to "dsop dhe abchRR in uneei hnur." Breed from this. In another fifty rounds or so we arrive triumphantly at our target "Drop the anchor in one hour."

The above argument in its pure form can be convincing only to persons already convinced. It asserts a functional difference between two nonsensical strings of letters. No person, or machine for that matter, looking for a sentence would notice a difference between "bsqm dshcbbbk,RR .nsurlei aknex" and "bsqp dshcbbbk,RR .nsurlei aknux." It is only because the believer has a distant goal in mind that he or she chooses one nonsense character string over the other. In the believers' argument the analogy of proteins to language is implicitly abandoned in the first rounds of the monkey's typing, since the character string does not have to obey any rules of spelling or grammar. The analogy to language is used simply to try to impress the unwary with the apparent production of sense from nonsense. My students in "Popular Arguments on Evolution" were uneasy with this argument when they read it in Dawkins's book, but they could not refute it. It is not easy for the casual reader to see that the illusion of steady, gradual evolution to a functional sentence is produced by an intellect, either the believer's directly or in some cases a computer program written by him, guiding the result to a distant goal. This of course is the antithesis of Darwinian evolution.

But perhaps there is a middle ground between the skeptic's insistence on absolute grammatical correctness and the believer's abandonment of grammatical rules. Suppose we allowed the vowels in the sentence to vary to produce something like "Drep tha enchir on une hoir." Such a sentence could probably still be recognized by someone, perhaps a sailor, even though all the words are misspelled. Or, alternatively, suppose we vary some consonants: "Trof tte ankhow im ode hous." Clearly some misspelled words would be easier to recognize than others and some letter substitutions (*t* for *d*, *k* for *c*) would be easier to follow than others (*r* for *t*, *l* for *g*). The ability of a sentence like that to function would depend a lot on the reader and the context.

To put this back into a protein context, it might be possible for a protein to tolerate a lot of amino acid substitutions and remain functional. (Again, when talking about proteins, *functional* means folded to a discrete, stable structure.) And in fact it has been known for a long time that this is true. Analogous proteins from different species—for example, human hemoglobin and horse hemoglobin—have differences between their amino acid sequences, yet fold to discrete and closely similar structures.

But what is the limit to tolerance for amino acid changes? Are proteins significantly more tolerant to changes in "spelling" than words are? Is there a point at which, like our sentences above, further changes will render a protein nonfunctional? What then is the probability of finding *some* member of a particular class in a reasonable time in a nondirected search? These are empirical questions and, although they can be speculated upon in the absence of relevant data, such speculations must be radically curtailed when data are available. A direct approach to the question, "How isolated are functional protein sequences?" would have been experimentally impossible twenty years ago, before the molecular biological revolution. But since the development of powerful tools to probe the molecules of life, an answer to that question appears to be within reach. Progress in this area is the topic of the following sections.

How Rare are Functional Proteins?

The observation that analogous proteins from different species could differ from each other, often by quite a bit, and yet retain the same compact shape led workers in the field to speculate that perhaps the exact identity of an amino acid at a particular position in a protein was not so important as its overall chemical properties. So, for example, if one finds an *I* at position 10 of hedgehog hemoglobin and an *L* in position 10 of the analogous protein from skunk, then perhaps the important feature is that both *I* and *L* prefer an oily environment, and maybe any other amino acid, such as *W*, *F*, or *V*, that prefers a similar environment would also be suitable at that position. This is something like saying that in a language perhaps all of the vowels are interchangeable. Taking the idea further, perhaps amino acids, such as S, A, H, and T, that prefer a watery environment could form an interchangeable group, and perhaps charged amino acids (E, D, R, and K) another group.

Fifteen years ago a man named Hubert Yockey published an article in the *Journal of Theoretical Biology*[1] showing that these considerations could enormously reduce the odds against finding a functional protein by trial and error. If we do not insist on the perfect diction of the typical skeptic, but allow some slurred speech in proteins, then the probability of finding a small, functional protein of one hundred amino acids in length is reduced from one in ten to the 130th power to one in ten to the 65th power—a reduction of sixty-five orders of magnitude! Yockey went on to show in the article that his calculation of one in 10^{65}, which he obtained from theoretical considerations, fit very closely with the number that could be calculated from considerations of the known sequence variability of the protein cytochrome c among many different species.

Now, the problem with Yockey's calculation for a believer in the sufficiency of natural law is that, although 10^{65} is enormously smaller than 10^{130}, it still is quite a large number. It has been calculated that there are about 10^{65} atoms in a galaxy. Thus, if Yockey was correct, the odds of finding a functional protein are about the same as finding one particular atom in the Milky Way. Not too likely. Well, if you were a believer, how might you answer this challenge? One way is through obfuscation, like the production of sentences from nonsense character strings, as was discussed above. A second way is by claiming that Yockey's calculation is inaccurate and that the known sequences of cytochrome c that he used to buttress his work do not reflect all the possible sequences that could produce a folded protein. The best way, though, in the absence of relevant data, is to produce your own calculation, starting from a separate set of independent principles, and show that the odds are not quite so long as Yockey thought. This is what has been done in an elegant series of calculations from the laboratory of Ken Dill[2,3] at the University of California at San Francisco.

Dill's laboratory asked a question that can be paraphrased as follows. Given a ten-by-ten square matrix (like a big checkerboard) and a string of pearls containing both black beads and white beads, in how many ways can a string of one hundred pearls be laid on the checkerboard so that each square contains one and only one pearl, and most of the black pearls are in the middle spaces of the board? This analogy is intended to represent a folding protein comprised of two types of amino acids, ones that prefer watery surroundings and ones that do not. After feeding this scenario into a computer, Dill's group obtained the surprising result that it wasn't that hard to fit the pearl necklace on the checkerboard in the right way. They then mathematically extrapolated from the two dimensional checkerboard to three dimensional space, and finally arrived at the conclusion that about one in 10^{10} amino acid sequences would yield a folded protein. That is a much smaller number than Yockey's (the federal government spends 10^{10} dollars, ten billion dollars, every three days) and brings the spontaneous generation of functional proteins into the realm of the credible.

The problem for a skeptic is how to refute Dill's calculation. It isn't easy, since few people are as mathematically talented as he and since it's hard to disprove the simplifying assumptions his model contains. Skeptics are free to criticize the assumptions, but there is enough uncertainty in such things to allow believers to tout Dill's calculation credibly over Yockey's. To resolve this dilemma, to gain firm ground to stand on, hard experimental results are required. Fortunately in the past several years such results have been forthcoming from the laboratory of Robert Sauer[4,5,6] in the department of biology at the Massachusetts Institute of Technology. We now turn to those crucial experiments.

Functional Proteins Are Very Rare

In the past twenty years the science of molecular biology has made enormous strides. It is now possible, in laboratories with such expertise, to cut up a gene, rearrange it to suit yourself, and place it back in a functioning biological system. Since genes code for proteins, one can also produce proteins made-to-order in this manner. Sauer's laboratory, in order to answer questions about protein structure that interested them, took the genes for several viral proteins, systematically took out small pieces of them (corresponding to instructions for three amino acids at a time), and inserted altered pieces back in the genes. They did this, three amino acids "codons" at a time, for the whole length of the gene. By clever manipulation of the altered pieces they were able to screen codons for all twenty amino acids at each position of the protein. This is like trying all twenty-six letters of the alphabet in turn at each position of a word. The altered genes were then placed in bacteria, which read the DNA code and produced chains of amino acids from them. It turns out that bacteria quickly destroy proteins that are not folded, so Sauer's group looked for the altered proteins that were not destroyed. By determining their sequences they could tell which amino acids in a given position were compatible with producing a folded, functional protein.

What did they see? In some positions of the protein, Sauer's group saw that a great deal of amino acid diversity could be tolerated. Up to fifteen of the twenty amino acids could occur at some positions and still yield a functional, folded protein. At other positions in the amino acid sequence, however, very little diversity could be tolerated. Many positions could accommodate only three or four different amino acids. Other positions had an absolute requirement for a particular amino acid; this means that if, say, a *P* does not appear at position 78 of a given protein, the protein will not fold *regardless of the proximity of the rest of the sequence to the natural protein.* In terms of our sentence analogy, this is like saying that, yes, all vowels are interchangeable, but that if the last *r* is changed to any other letter, such as *s* ("Drop the anchor in one hous"), the protein sentence is no longer understandable.

Sauer's results can be used to calculate the probability of finding a given protein structure.[6] We proceed in the following manner. If any of ten amino acids can appear in the first position of a given functional protein sequence, then the odds are one in 2 that a nondirected search will place one of the allowed group there. If any of four amino acids can appear in the second position, then the odds are one in 5 of finding one of that group, and the odds of finding the correct amino acids next to each other in the first two positions are one-half times one-fifth, which is one-tenth. Suppose in the third position there is an absolute requirement for *G*. Then the odds of getting a *G* at that position are one in twenty and the odds of getting the first three amino acids right are now up to one in two hundred. In this aspect it is like winning a trifecta in horse racing. Over the course of one hundred amino acids in our small protein, the odds quickly reach astronomical numbers.

From the actual experimental results of Sauer's group it can easily be calculated that the odds of finding a folded protein are about one in 10 to the 65th power.[6] To put this fantastic number in perspective, imagine that someone hid a grain of sand, marked with a tiny *X*, somewhere in the Sahara Desert. After wandering blindfolded for several years in the desert you reach down, pick up a grain of sand, take off your blindfold, and find it has a tiny *X*. Suspicious, you give the grain of sand to someone to hide again, again you wander blindfolded into the desert, bend down, and the grain you pick up again has an X. A third time you repeat this action and a third time you find the marked grain. The odds of finding that marked grain of sand in the Sahara Desert three times in a row are about the same as finding one new functional protein structure. Rather than accept the result as a lucky coincidence, most people would be certain that the game had been fixed.

The number of one in 10^{65}, arrived at by Sauer's experimental route, is virtually identical to the results obtained by Yockey's theoretical calculation and his deduction from natural cytochrome c sequences! It therefore strongly reinforces our confidence that a correct result has been obtained. Sauer's group obtained closely similar results for two different proteins: arc repressor[4] and lambda repressor.[5,6] This means that all proteins that have been examined to date, either experimentally or by comparison of analogous sequences from different species, have been seen to be surrounded by an almost infinitely wide chasm of unfolded, nonfunctional, useless protein sequences. There are no ledges, no buttes, no steppingstones to cross the chasm.

The conclusion that a reasonable person draws from this is that the laws of nature are insufficient to produce functional proteins and, therefore, functional proteins have not been produced through a nondirected search.

Implications of Protein Sequence Isolation

The numerical concreteness of Sauer's and Yockey's results is breathtaking. When a skeptic sees a drawing of *Mesonychid* next to the Zeuglodon whale, he or she intuitively realizes that the transformation is highly improbable. But how improbable? There is no way to put a quantitative measure on the difference between a doglike animal and a whale, and believers in the relentless application of physical law take advantage of this by verbally minimizing the differences.

The situation is otherwise with proteins. Because there is a discrete set of amino acids and a finite number of positions in a given protein, the odds of attaining a folded, functional protein can be calculated quite closely, but only if the tolerance of proteins to amino acid substitution is known. Thanks to Sauer and Yockey we now have such quantitative data.

It is important to realize that Sauer's and Yockey's results hold *whether or not the system can replicate and is subject to Darwinian selection*. The odds against finding a new functional protein structure remain astronomical in either case. This is because Darwinian selection can only discriminate based on function and, with the exception of those found in living organisms, virtually all protein sequences are functionless. An amino acid sequence can be replicated and mutated in living organisms till the cows come home, and the odds are still one in 10^{65} that a new functional protein class will be produced.

The problem of the isolation of functional protein sequences is a vivid illustration of the truth of the symposium thesis,

> Darwinism and neo-Darwinism as generally held and taught in our society carry with them an *a priori* commitment to metaphysical naturalism, which is essential to make a convincing case in their behalf.

The skeptic can accept Sauer's and Yockey's results with equanimity because his world is not necessarily limited to those phenomena that can be explained by naturalism. Furthermore, the skeptic can happily concede that many biological phenomena *are* explained by natural laws. He can agree that beak shape and wing color can change under selective pressure, or that different proteins in the same structural class, such as the alpha and beta chains of hemoglobin, may have arisen through Darwinistic mechanisms. But the believer in the universal application of physical law is stuck. He must maintain, *against the evidence*, that different protein classes, like cytochromes and immunoglobulins, found their way by raw luck through the vast, dark sea of nonfunctional sequences to the tiny islands of function we observe experimentally. He must maintain, *without any evidence*, that *Mesonychid* gave birth over time to the whale. And why, we ask, must he maintain these positions against impossible odds and without supporting evidence? Because, he replies, I can measure only material phenomena, and therefore nothing else exists.

In closing I would like to paraphrase Hubert Yockey,[7] who in his career repeatedly pointed out facts that are not supposed to be mentioned in polite scientific company: "Since science has not the vaguest idea how [proteins] originated, it would only be honest to admit this to students, [to] the agencies funding research, and [to] the public."

NOTES

[1]Yockey, H. P. (1978), "A Calculation of the Probability of Spontaneous Biogenesis by Information Theory," *Journal of Theoretical Biology* 67:377-398.

[2]Lau, K. F., & Dill, K. A. (1989), "A Lattice Statistical Mechanics Model of the Conformational and Sequence Spaces of Proteins," *Macromolecules* 22:3986-3994.

[3]Chan, H. S., & Dill, K. A. (1990), "Origins of Structure in Globular Proteins," *Proceedings of the Natural Academy of Sciences USA* 87:6388-6392.

[4]Bowie, J. U., & Sauer, R. T. (1989), "Identifying Determinants of Folding and Activity for a Protein of Unknown Structure," *Proceedings of the National Academy of Sciences USA* 86:2152-2156.

[5]Bowie, J. U., Reidhaar-Olson, J. F., Lim, W. A., & Sauer, R. T. (1990), "Deciphering the Message in Protein Sequences: Tolerance to Amino Acid Substitution," *Science* 247:1306-1310.

[6]Reidhaar-Olson, J. F., & Sauer, R. T. (1990), "Functionally Acceptable Substitutions in Two α-Helical Regions of λ Repressor," *Proteins: Structure, Function, and Genetics* 7:306-316.

[7]Yockey, H. P. (1981), "Self Organization Origin of Life Scenarios and Information Theory," *Journal of Theoretical Biology* 91:13-31.

6a

Response to Michael J. Behe
The Process, Described Properly,
Generates Complexity in Good Time
Leslie K. Johnson

Abstract: Dr. Behe argues that a protein performing a given function in the complex environment of the cell is such an improbable thing that it could not be expected to arise in the time span available on earth. The problem with his formulation is this: the process he models is not the same process described by the theory of evolution. Evolution requires inheritance, mutation, and selection. Dr. Behe's process involves only inheritance and mutation. Once you have a simple replicating structure (inheritance) that from time to time suffers changes in its replication code (mutation), *and* particular mutants arise that out-multiply others (selection), then the mutant type becomes common, forming the background population in which the next winning mutation occurs. In this way, each stepwise "gain" (in light of the final result) is consolidated.

A PROTEIN HAS BEEN PRESENTED as a complex thing. It is. There are limited ways it can be modified and still function in the cell. That is true. The exact ways a particular protein can differ without destroying function have been investigated experimentally with exquisite technique. A protein is in essence a chain of discrete beads or elements of finite number and of describable relative availability for stringing. Therefore all possible ways of randomly constructing a chain of equivalent length can be simply calculated. The elements in a protein chain are viewed as steps that have to have occurred.

In such a model, with a chain of any appreciable length, and an amino acid soup of any appreciable diversity, the probability of getting one of the few possible chains that "work" quickly gets exceedingly small, so small, that for our minds to grasp the unlikelihood, we must resort to metaphor. All this is true.

The process Professor Behe describes—a process of stepwise

amino acid substitutions adding up to an improbable product; a process extended in time but with a probability of occurrence analyzed no differently than had it all been assembled in "one fell swoop"—is not analogous to the process of evolution by natural selection. Yes, like organic evolution, there are replication and mutation. But what has been left out are the filters, the sieves that at every generation sift the outcomes. The sieve is natural selection. No discerning selector is implied.

Selection is a way of describing the fact that, in the environment in question, some of the variants will be more successful than others in populating the next generation with their sort. These variants are better at lasting long enough to make copies, and better at making relatively many of these copies. No selector is implied, but "sense" does build itself into the process. Which variants do relatively well is not entirely haphazard. On average, successful variants surmount the complex challenges of their environment by happening to be a bit more complex themselves in the effective sorts of ways.

As this mechanical process is iterated, and variants of differing success continue to pop up, the diversity in the total collection rises. Rising diversity means that the environment in which the variants exist and replicate gets more complex over time. So, yet more complex ways of existing and replicating are the ones that work relatively better in later generations. Viewed overall, the unfolding scenario has the look of progress.

The analogy between typing monkeys and evolution has a flaw, which is teleology. Teleology is a goal toward which something is working. In the monkey example, the goal is the character string that spells "Drop the anchor in one hour." The monkey types character strings of lengths similar to the goal. Every time the random product gets the same letter in the same place as the goal, that character is inserted in that site with each succeeding string of letters the monkey types. Naturally, by and by, the goal is reached. The teleology is not in the mind of the monkey, it is true, but is present because the game is rigged.

A little less teleological is the transmogrification of everyday food preparation into a practical, delicious showpiece of regional cuisine. Night after night, throughout the region, meals are prepared. Haphazard elements affect the product: what's in season, what's on hand, what's convenient at the time. Children poke at it, husbands mumble over it, but once in a while someone says, "Hey, that's delicious—write it down!" A recipe appears. The recipe gets replicated whenever a guest or a relative asks to have it, and it is replicated even more when it is included in the PTA fundraiser cookbook. Each new owner of the recipe is likely to alter it a bit, leaving out a disliked ingredient, adding a radish rosette. New environments affect what is made: microwave ovens, say, or the Surgeon General's recommendations. A recipe that is really successful in leaving descendants bears a name everyone recognizes— fajitas, ginger beer, bubble-and-squeak.

So, with somewhat accidental variation, "filters" that operate every time the dish is made, and replication, we have an outcome: a regional dish that could not have been specified at the outset in the cabins of the first local settlers. The analogy, however is flawed. Design does creep in. Food preparers do think, and have short-term goals in mind.

Other analogies avoid the problem of teleology. You and I are the highly improbable outcomes of all the chance meetings, feelings of love, mutual attractions, rapine roughness, release of particular ova, and plain old fluid dynamics of all the couplings of all our ancestors since the dawn of history. We were not envisioned in our glorious uniqueness by any of the players in our past. But this analogy, too, is imperfect. We are, arguably, no more complex than our ancestors in Mesopotamia, or wherever.

It is Tom Ray's computer program that makes the best analogy I know of to the process of organic evolution. The elements of replication, production of new variation, and non-teleological, automatic selection are present. These elements produce novelty, complexity, diversity.

The best example, of course, is the real thing: organisms surviving and reproducing in environments in which some types do better than others. Successful variants tend to be those good at acquiring whatever the needed resources are, converting them efficiently into growth and offspring, lasting long enough to do so, and helping organisms with genotypes most like one's own. For those wanting to understand what evolutionary biologists mean by evolution, organismal biology merits careful study.

To touch on something else, the production of new variants is sometimes equated with point mutation. A point mutation is an altered nucleotide in the genetic material. An analogy to this is a substitution in a typed character string. When evolutionary biologists speak of mutation, they mean point mutation and more. Mutations are spontaneous gene changes, including point mutations at one or several nucleotides, changes in chromosome number or structure, and shuffling of parts of genes, as, for example, transposition of gene segments.

All this becomes significant when we seek to understand evolutionary attainment in groups as different as bacteria, fungi, green plants, and mammals. Biochemically, it looks as if all life started from one basic kind a long time ago. During diversification, rather different modes of organization were achieved, such as unicellularity, cellular differentiation, or development that proceeds by induction. These modes of organization put constraints on what further kinds of innovation were likely to occur.

Evolution in bacteria, for example, tends to involve minor changes in the code, RNA, which in turn affects metabolic pathways. Flowering plants are developmentally simple and morphologically plastic, and often speciate by multiplication of chromosome number. They are essentially constrained from evolving nervous systems by the cellulose walls that enclose each cell. Mammals have complex, interactive development. Their evolution frequently involves regulatory genes that affect developmental timing and differential sensitivity of different parts of the neuroendocrine system. A small difference early leads to a big difference in adult structure and function.

This means that evolution can be expected to occur with differing tempo and mode at different times during the history of life and in different taxonomic groups. As we learn more and more about molecular genetics and developmental biology, we can make more and more refined predictions about which groups are likely to speciate a lot and under what circumstances, and what sorts of novelty will appear in the daughter species. Deepened understanding will permit new tests of the validity of the theory.

Darwinism has met the challenge of the explosion of new information generated by the growth of molecular biology, and is becoming integrated with it in ways that get richer with the passage of each publishing day. The theory is healthy.

True, one can find practicing scientists who are skeptical about evolution. Without having conducted a survey, I will brazenly hypothesize that such skeptics will be drawn disproportionately from technology fields and fields that focus on more physicochemical levels of organization. These fields have principles of organization of their own which need not be much perturbed by the parade of life. Such principles include quantum mechanics or electron orbital theory.

The big theory for biologists, however, especially those who work at the most emergent levels of organization (such as social behavior), is evolution by natural selection. As an organizing principle that is bolstered by, tested against, and modified according to evidence, it has tremendous explanatory power.

Take one small set of biologists, those who work on amphibians, a minor group of animals. Since 1970, amphibian biologists have been producing more than 1,000 titles per year, according to the Zoological Record. Topics include vocalization, larval traits, endocrinology, the fossil record, reproductive strategies, development, the musculoskeletal system, sensory reception, molecular evolution, cytogenetics, biogeography, and digestion. William Duellman and Linda Trueb produced a big new book, *The Biology of Amphibians*. The framework into which they fit all this stuff is evolution. This would be true as well if they made an Encyclopedia of Amphibians.

With evolution as an organizing scheme, such an encyclopedia would be compelling and understandable. Without evolution, it would be as exciting as a fourteen-volume set of urban telephone books.

This is why evolution works for me and for my fellow biologists.

6b
Reply to Leslie K. Johnson
Michael J. Behe

FAJITAS? GINGER BEER? Bubble and Squeak? Is this the reply that Darwin's vaunted theory gives to a serious, quantitative, detailed, experimental challenge? Dr. Leslie Johnson asserts that "the theory is healthy," but the replies it gives to probing questions are those of a ninety-eight pound weakling.

I am very pleased that Dr. Johnson agrees with me that the example of the monkey producing a functional sentence by replacing a letter at a time in a nonsense character string is illegitimate. She appears not to realize, however, that it is Darwinians who have advanced this example. Perhaps she can inform her co-panelist Michael Ruse, who uses a similar analogy in his book *Darwinism Defended*, of his error. And perhaps he can then contact Richard Dawkins, who uses the analogy in *The Blind Watchmaker*, to tell him of their mutual mistake.

The book that launched Darwin's theory was entitled *The Origin of Species*. Darwinism's appeal rests largely on its claim to be able to explain the origin of the great complexity of the biological world, a complexity that all admit gives the appearance of design, without recourse to non-natural agents. But when detailed questions are asked about the origin of biological structures, proponents of the theory all too frequently resort to hand-waving and metaphor of the kind Dr. Johnson offers. For example, we are told by her that "we seek to understand evolutionary attainment . . . ," "evolution in bacteria tends to involve minor changes . . . ," and "[mammals'] evolution frequently involves regulatory genes." Regrettably, however, Dr. Johnson never gets around to telling us, even for a single example, exactly which evolutionary changes gave rise to which biological structures in the real world. We are thus left wondering how she knows that organisms have evolved at all.

Dr. Johnson is not alone in her style of argumentation: *no one* at this conference has argued the merits of Darwinism by pointing to a complex biological structure and explaining in detail how it arose from a simpler structure through the agency of natural selection. Instead we are implicitly invited to imagine such developments by means of fuzzy mental images, playing horror movie-like transmogrifications in our minds. This is the appeal of much of the "computer evolution" work that Dr. Johnson cites favorably: images can "evolve" like Dr. Jekyll on the computer screen without having to be tested for their ability to function in the real world.

But, then, if no one actually uses Darwin's theory to give plausible, detailed explanations for the origin of complex biological structures, what exactly is it good for? To use as a "framework," Dr. Johnson tells us. "Without evolution" descriptions of nature "would be as exciting as . . . telephone books." That may be true for Dr. Johnson, but it is not true for children visiting a zoo, it is not true for most laypersons, and it wasn't true for pre-Darwinian biologists like Linnaeus and Cuvier. It is a dangerous intellectual game to confuse one's own mental filing cabinets for the real world.

7
The Incompleteness of
Scientific Naturalism
William A. Dembski

FIRST LET ME EXPRESS my thanks to the organizers of this symposium for the opportunity to present certain ideas that for some time now have exercised me. The occasion for this symposium is Phillip Johnson's book *Darwin on Trial.* The title would suggest that Johnson's main concern is with Darwinism and neo-Darwinism proper. Nevertheless, I would claim that Johnson's book is as much about a philosophical world view used to prop up Darwinism as it is about Darwinism. Atheism, materialism, scientism, and secular humanism are a few of the names attached to this world view. Yet the name I like best and find most descriptive is *scientific naturalism.*

I want here to examine scientific naturalism. I am going to argue that this view has a serious defect—it is *incomplete.* As a consequence of this defect I shall argue that it is legitimate within scientific discourse to entertain questions about supernatural design. The backdrop for this discussion will comprise two areas in mathematics: computational complexity theory and probability theory.

First let's be clear what we mean by scientific naturalism. The key ingredient in scientific naturalism is, let me say it, *naturalism.* Naturalism as a world view has two components: (1) It is a meta-physical doctrine about what things exist in the world. These include material objects and sometimes (as for the philosopher Willard Quine) mathematical objects such as sets. Excluded are supernatural beings, nonmaterial interventions, divine meddlings, etc. (2) Naturalism includes an epistemological doctrine about how the things permitted under this metaphysical doctrine are to be explained—i.e., they are to be explained naturalistically. I am not sure that naturalistic explanation is a perfectly clear notion, but what is clear is that naturalistic explanation excludes any sort of appeal to nonmaterial intervention, divine meddling, etc.

Where does the "scientific" in scientific naturalism come in? As a world view, scientific naturalism regards itself as continuous with

79

science. It therefore looks to our scientific understanding of the world for its justification. This last point distinguishes scientific naturalism from naturalism simpliciter. It is also this last point that is responsible for scientific naturalism being incomplete.

To see what is at stake let me quote the last line of Edwin Hubble's *The Realm of the Nebulae*: "Not until the empirical resources are exhausted need we pass on to the dreamy realms of speculation." When Hubble wrote that line in the 1930s, he clearly believed that the empirical resources would not be exhausted and that our entrance into the dreamy realms of speculation could be postponed indefinitely.

Against this I would argue that empirical resources come in limited supplies and do get exhausted. Moreover, as soon as empirical resources are exhausted, naturalism can no longer find its justification in science. This then is the incompleteness of scientific naturalism, namely, the incapacity of science to justify naturalism once the empirical resources wherewith science limits itself get exhausted.

Next I want to focus on two empirical resources, one computational, the other probabilistic. I want to show how even the possibility of these resources being exhausted undermines the completeness of scientific naturalism—the pretension, as far as I'm concerned, that a complete understanding of the world is possible apart from God. Since this talk is addressed primarily to non-mathematicians, I'll begin by considering the words of a well-known American philosopher, Woody Allen.

Woody Allen probably didn't think that God would take him seriously when he quipped,

> If only God would give me some clear sign! Like making a large deposit in my name at a Swiss bank.[1]

But what if God had taken Allen seriously? Would an unexpected $7,000,000, say, in Allen's Swiss bank account have convinced him that God was real? Suppose that a thorough examination of the bank records failed to explain how the money appeared in Allen's account. Should Allen have inferred that God had given him a sign?

Since I can't answer for Allen, let me answer for myself. If I were a famous personality having uttered Allen's remark and subsequently found an additional $7,000,000 in my Swiss bank account, I would certainly not have attributed my unexpected good fortune to the largesse of an eccentric deity. It's not that I don't believe in God. I do. But my theology constrains me to think it unworthy of God to grant flippant requests like Allen's and then apparently ignore the urgent requests of so many suffering people in the world.

I would refuse to acknowledge a miracle for theological reasons. Barring theological reasons, however, I would still refuse to acknowledge a miracle. Why? Well, other explanations readily come to mind. If I had uttered the remark and were as famous as Allen, and if $7,000,000 had appeared in my account, I would probably have concluded that some eccentric billionaire with a religious agenda was trying to convert me to his cause. The strange appearance of the $7,000,000 would have been fiendishly designed to make me believe in God. But alas, I was too clever for them.

There is a point to these musings. Allen's remark is clearly funny; however, if taken seriously it is self-defeating. If God were in fact to do what Allen requested, Allen and just about anyone else would remain unconvinced. The question therefore arises whether God can do anything, either in response to a request like Allen's or otherwise, which would provide convincing proof that he and no one else had acted.

Let's put it this way: Is there anything that has, could, or might happen in the world from which it would be reasonable to conclude that God had acted? Are there or could there be any facts in the world for which an appeal to God is the best explanation? Or to reverse the question, is God always an easy way out, a lame excuse, a prescientific device that invariably misses the best explanation?[2]

We are asking a transcendental question in the Kantian sense: What are the conditions for the possibility of discovering design (i.e., supernatural intervention, nonmaterial interference, divine meddling, call it what you will) in the actual world? This question must be answered at the outset, for if this world is the type of place where anything even in principle that happens can be adequately explained apart from teleology and design, then it makes no sense to look for design in what actually happens. Might the world do something, however quirky, that would convince us of design?

An illustration might help. Imagine a peculiar art studio comprised of ten-inch by ten-inch canvases, a full range of oil paints, and a robot that paints the canvases with the paints. In painting the canvases, the robot divides each canvas into a ten by ten grid of one-inch squares, and paints each square with precisely one color. Imagine that this robot also has visual sensors and thus can paint scenes presented to its visual field, though only crudely, given the coarse-grained approach it adopts to painting.

Imagine next that Elvis and an Elvis impersonator come to have their portraits painted by this robot. Will the portraits distinguish Elvis from his impersonator? Because the representations on canvas are so crude, if the impersonator is worth his salt, the two portraits will be indistinguishable. Our imaginary art studio cannot distinguish the real Elvis from the fake Elvis.

This example indicates what is at stake in determining whether design has at least the possibility of being detected and empirically grounded. Putative instances of design abound. But is it possible within this world to distinguish authentic from spurious design should instances of authentic design even exist? Or is this world like the art studio? Just as the portraits painted at the studio cannot distinguish the real from the fake Elvis, so too is it impossible for our empirical investigations of the world to distinguish authentic from spurious design?

Scientific naturalism prefers to think just this, namely, that the world is the kind of place where all objective phenomena can be explained by purely naturalistic factors. Non-naturalistic factors therefore become not only redundant but also illegitimate to explanation. As George Gaylord Simpson put it,

> There is neither need nor excuse for postulation of nonmaterial intervention in the origin of life, the rise of man, or any other part of the long history of the material cosmos.[3]

Simpson claims that the world is the kind of place where no objective, empirical finding can ever legitimately lead us to postulate design (what he calls "nonmaterial intervention").

That is a bold claim. The question remains whether it is true. In the case of the art studio, it is true that robot portraits of Elvis and his impersonator will fail to distinguish the two. The paintings produced by the studio are simply too coarse grained to do any better. From these paintings there is, to use Simpson's phrase, "neither need nor excuse for postulation of" two Elvises, the real and the fake. From the portraits alone we might legitimately infer only one sitter. But is the world so coarse grained that it cannot even in principle produce events that would evidence design? That is what Simpson seems to be affirming. A little reflection, however, indicates that this claim cannot be right.

We consider a thought experiment, one I call "The Incredible Talking Pulsar." Imagine that astronomers have discovered a pulsar some three billion light years from the earth. The pulsar is, say, a rotating neutron star that emits regular pulses of electromagnetic radiation in the radio frequency range. The astronomers who found the star are at first unimpressed by their discovery. It's only another star to catalogue. One of the astronomers, however, is a ham radio operator. Looking over the pattern of pulses one day, he finds that they are in Morse code. Still more surprisingly, he finds that the pattern of pulses signals English messages in Morse code.[4]

Word quickly spreads within the scientific community, and from there to the world at large. Radio observatories around the globe start monitoring the "talking" pulsar. The pulsar isn't just transmitting random English messages, but is instead intelligently communicating with the inhabitants of earth. In fact, once the pulsar has gained our attention, it identifies itself. The pulsar informs us that it is the mouthpiece of Yahweh, the God of both the Old and the New Testaments, the creator of the universe, the final judge of humankind.

Pretty heady stuff you say. But to confirm this otherwise extravagant claim, the pulsar agrees to answer any questions we might put to it. The pulsar specifies the following method of posing and answering questions. The descendants of Levi are to make an ark like the one originally constructed under Moses (see Exodus 25). This ark is to be placed on Mount Zion in Israel. Every hour on the hour a question written in English is to be placed inside the ark. Ten minutes later the pattern of pulses reaching earth from the pulsar will answer that question, the answer being framed as an English message in Morse code.[5]

The information transmitted through the pulsar proves to be nothing short of fantastic. Medical doctors learn how to cure AIDS, cancer, and a host of other diseases. Archaeologists learn where to dig for lost civilizations and how to make sense out of them. Physicists get their long-sought-after unification of the forces of nature. Meteorologists are forewarned of natural disasters and weather patterns years before they occur. Ecologists learn effective methods for cleansing and preserving the earth. Mathematicians obtain proofs to many long-standing open problems—in some cases proofs they can check, but proofs they could never have produced on their own. The list of credits could be continued, but let us stop here.

What shall we make of the pulsar? Whether the pulsar is in fact the mouthpiece of Yahweh, the pulsar creates serious difficulties for scientific naturalism. Not only is there no way to square the pulsar's behavior with our current scientific understanding of the world, but it is hard to conceive how any naturalistic explanation will ever account for the pulsar's behavior. For instance, our current scientific understanding based on Einsteinian special relativity tells us that messages cannot be relayed at superluminal speeds. Since the pulsar is three billion light years from the earth, any signal we receive from the pulsar was sent billions of years ago. Yet the pulsar is "responding" to our questions within ten minutes of the written questions being placed inside the ark. The pulsar's answers therefore seem to precede our questions by billions of years.

To get around this, scientific naturalists might want to postulate reverse causality or superluminal signaling. Naturalists might find this idea more congenial than postulating "nonmaterial intervention," but reverse causality and superluminal signaling do not even begin to address the questions raised by the pulsar. It is inescapable that in dealing with the pulsar we are dealing with not just an intelligence, but with a super-intelligence. Now by a super-intelligence I don't mean an intelligence that at this time surpasses human capability, but which in time humans can hope to attain. Nor do I mean a super-human intelligence that might nevertheless be realized in some finite rational material agent embedded in the world (say an extraterrestrial intelligence or a conscious super-computer). By a super-intelligence I mean a supernatural intelligence, i.e., an intelligence surpassing anything that physical processes are capable of offering. This intelligence exceeds anything that humans or finite rational agents in the universe are capable of, even in principle.

How can we see that the pulsar instantiates a super-intelligence? The place to look is computer science. There are problems in computer science that can be proven mathematically to require more computational resources for their solution than are available in the universe. Think of it this way. There are estimated to be no more than 10^{80} elementary particles in the universe. The properties of matter are such that circuits cannot be switched faster than 10^{45} times per second.[6] The universe itself is about a billion times younger than 10^{25} seconds (assuming that the universe is at least a billion years old). Given these upper bounds we can confidently assert that no computation exceeding $10^{80} \times 10^{45} \times 10^{25} = 10^{150}$ elementary steps is possible within the universe. By an elementary step I mean the switching of a two-state device, conceived abstractly as the switching of a binary integer (= bit). For a computation of this complexity therefore to be carried out in the universe, every available elementary particle in the universe would have to serve as an elementary storage device (= memory bit) capable of switching at 10^{45} hertz over a period of a billion billion years.

10^{50} is incredibly generous as an upper bound on the complexity of computations possible in the universe. Here are a few reasons why a much smaller bound will do: (1) quantum mechanical considerations indicate that reliable memory storage is unworkable below the atomic level[7] since at this level quantum indeterminacy will make not only storage, but also reading and writing of information impossible. Hence each elementary storage device will have to consist of more than one elementary particle. (2) The preceding calculation treats the universe as a giant piece of random access memory that is controlled by a processor outside the universe operating at 10^{45} hertz with instant access to any memory location in RAM. In fact, the processor will itself have to take up part of the universe. Moreover, its access to memory locations will have in some cases to be measured in light years and not in 10^{45} second chunks. Even with massively parallel processing, computation speeds will fall far below the 10^{45} hertz upper bound. (3) Finally, the bound of 10^{25} seconds for the maximum running time of a computation is excessive since the heat death of the universe will probably have occurred by then. Suffice it to say, even with the entire universe functioning as a computer, no computation requiring 10^{150} elementary steps, much less 10^{150} floating point operations, is feasible.

Now it is possible to pose problems in computer science for which the quickest solution requires well beyond this number steps, yet for which with a solution in hand it is possible even for humans using ordinary electronic computers to check whether the solution is correct. Factoring integers into primes is thought to be one such problem. Since the factorization problem is easy to understand, let me treat it as though it were one of the "provably hard problems." If at some time in the future a "quick" algorithm is found for factoring numbers, we shall need to modify this example; nevertheless, our contention that there are problems whose solution is beyond the computational resources of the universe, yet verifiable by humans, will still hold.[8]

What is the factorization into primes of 1961? Solving this requires a bit of work. But if you are given the prime numbers 37 and 53, it is a simple matter to check whether these are prime factors of 1961. In fact

37 x 53 = 1961. Factoring is hard, multiplication is easy. We can therefore go to our pulsar with numbers thousands of digits long and ask it to factor them. Factoring numbers that long is totally beyond our present capabilities and in all likelihood exceeds the computational limits inherent in the universe by many, many orders of magnitude. (When I was following the literature on factoring a few years back, numbers beyond two hundred digits in length could not be factored unless they had either small or special prime factors.) Nevertheless, it is easy enough to check whether the pulsar is getting the factorizations right, even for numbers thousands of digits in length.[9]

What lesson can we learn from the pulsar? I claim we should infer that a designer in the full sense of the word is communicating through the pulsar, i.e., a designer who is both intelligent and transcendent. Intelligence is certainly not a problem here. Alan Turing's famous test for intelligence pitted computer against human in a contest where a human judge was to decide which was the computer and which was the human.[10] If the human judge could not distinguish the computer from the human, Turing wanted intelligence attributed to the computer.

This operationalist approach to intelligence has since been questioned, by theists on one end and hard-core physicalists on the other. But the basic idea that there is no better test for intelligence than coherent natural language communication remains intact. If we cannot legitimately attribute intelligence to the pulsar, then no attribution of intelligence should count as legitimate. Transcendence is clear as well, given our discussion of intractable computational problems. Suffice it to say, a being that solves problems beyond the computational resources of the material world is not material. When we can confirm that such problems have in fact been solved for us, we cannot avoid postulating "nonmaterial intervention."

The pulsar demonstrates that ours is the type of world where design has at least the possibility of becoming perfectly evident—with the pulsar, empirical validation for design can be made as good as we like. In the actual world, design is therefore not only possible but also empirically knowable. I have belabored this point because it is a point scientific naturalism would rather not grant. Once, however, it is granted that the occurrence of certain events might require us to postulate "nonmaterial intervention," we need to consider whether any events that have actually occurred require us to postulate such intervention. It is obvious that the pulsar is an exercise in overkill. No instance of design so crashingly obvious is known. Science fiction has therefore done its work for us. It is time to put science fiction to rest, and look at what solid evidence there is for design in the actual world. We therefore leave computational resources and turn to probabilistic resources.

I use the term *probabilistic resources* to describe what I call replicational resources on the one hand, and specificational resources on the other. To appreciate what is at stake with these resources let us consider two examples. The first illustrates replicational resources, the second specificational resources.

Here is the first example. Imagine that a massive revision of the criminal justice system has taken place. Henceforward a convicted criminal is sentenced to serve time in prison until he flips n heads in a row, where n is selected according to the severity of the offense (we assume that all coin flips are fair and are duly recorded; no cheating is possible). Thus for a ten-year prison sentence, if we assume the prisoner can flip a coin once every five seconds (this seems reasonable), eight hours a day, six days a week, and given that the average attempt at getting a streak of heads before tails is 2 (= $\Sigma_{1 \leq i < \infty} i2^{-i}$), then he will *on average* attempt to get a string of n heads once every 10 seconds, or 6 attempts a minute, or 360 attempts an hour, or 2,880 attempts in an eight-hour work day, or 901,440 attempts a year (assuming a six-day work week), or approximately 9 million attempts in ten years. Nine million is approximately 2^{23}. Thus if we required of a prisoner that he flip 23 heads in a row before being released, we could expect to see him out in approximately ten years. Of course specific instances will vary—some prisoners being released after only a short stay, others never recording the elusive 23 heads!

Now consider the average prisoner's reaction after about ten years when he finally flips 23 heads in row. Is he shocked? Does he think a miracle has occurred? Absolutely not. Given his replicational resources, i.e., the number of opportunities he had for observing 23 heads in a row, he could expect to get out of prison in about ten years. There is in fact nothing improbable about his getting out of prison in this span of time. It is improbable that on any given occasion he will flip 23 heads in a row. But when all these occasions are considered jointly, it becomes quite probable that he will be out of prison within the ten years' time. The prisoner's replicational resources comprise the number of occasions he has to produce 23 heads in a row. If his life expectancy is better than ten years, he has a good chance of getting out of prison. In short, his replicational resources are adequate for getting out of prison.

If, however, the number of heads a prisoner must flip in a row is exorbitant, then his replicational resources will be inadequate for getting out of prison. Consider a prisoner sentenced to flip 100 heads in a row. The probability of getting 100 heads in a row on a given trial is so small that he has no practical hope of getting out of prison, even if his life expectancy was dramatically increased. If he could, for instance, make 10 billion attempts each year to obtain 100 heads in a row, then he stands only an even chance of getting out of jail in 10^{20} years. His replicational resources are so inadequate for obtaining the desired 100 heads that it's pointless to entertain hopes of freedom.[11]

With replicational resources the question is how many opportunities exist for observing a specific event (in the preceding example the event was flipping n heads in a row). With specificational resources the question is how many opportunities are there for specifying an as yet undetermined event. Lotteries provide the perfect vehicle for illustrating specificational resources. Indeed, each lottery ticket is a specification. To illustrate specificational resources, consider now the following lottery to end all lotteries: In the interest of eliminating the national deficit, the federal government agrees to hold a national lottery in which the grand prize is to be dictator of the United States for a single day— i.e., for twenty-four hours the winner will have full power over every aspect of government. If a white supremacist wins, he can order the wholesale execution of nonwhites. If a porn king wins, he can order this country turned into a giant debauch. If a pacifist wins, he can order the destruction of all our weapons . . . . The more moderate elements of the society will clearly want to prevent the loony fringe from winning, and will therefore be inclined to invest heavily in this lottery.

This natural inclination, however, is mitigated by the following consideration: the probability of any one ticket winning is 1 in 2^{100}, or approximately 1 in 10^{30}. To buy a ticket, the lottery player pays a fixed price and then records a 0-1 string of length 100—whichever string he chooses. He is permitted to purchase as many tickets as he wishes, subject only to his financial resources and the time it takes to record the 0-1 strings of length 100. The lottery is be drawn at a special meeting of the United States Senate: By alphabetical order each senator is to flip a coin once and record the resulting coin toss.

Suppose now that the fateful day has arrived. A trillion tickets have been sold at ten dollars apiece. To prevent cheating, Congress has enlisted the services of the National Academy of Sciences. Following the NAS's recommendation, each ticket holder's name is duly entered onto a secure data base, together with the tickets purchased and the ticket numbers (i.e., the bit strings relevant to deciding the winner). All this information is now in place. After much fanfare the senators start flipping their coins. As soon as Senator Zygmund has announced his toss, the data base is consulted to determine whether the lottery had a winner. Lo and behold, the lottery did indeed have a winner—Joe "Killdozer" Skinhead, leader of the White Trash Nation. Joe's first act as dictator is to raise a swastika over the Capitol.

From a probabilist's perspective there is one overriding implausibility to this example. The implausibility rests not with the federal government's sponsoring a lottery to eliminate the national debt, nor with the fascistic prize of being dictator for a day, nor with the way the lottery is decided at a special meeting of the Senate, nor even with the fantastically poor odds of winning the lottery. The implausibility rests with the lottery's having a winner. Indeed, as a probabilist myself, I would encourage the federal government to institute such a lottery if it could redress the national debt, for I am convinced that if the lottery is run fairly, there will be no winner. The odds are simply too much against it.

Suppose, for instance, that a trillion tickets are sold at ten dollars apiece (this would cover the deficit as it stands in 1992). What is the probability that one of those tickets (= specifications) will match the winning string of 0's and 1's drawn by the Senate? An elementary calculation shows that this probability can be no greater than 1 in 10^{18}. This is a tiny probability. Even if we increase the number of lottery tickets sold by several orders of magnitude, there still won't be enough sold for the lottery to stand a reasonable chance of having a winner. Since lottery tickets are specifications, this is equivalent to saying there aren't enough specifications to specify the event in question (i.e., the winning of the lottery).

Often it is necessary to consider replicational and specificational resources in tandem. Suppose for instance in the preceding lottery that the Senate will hold up to a thousand drawings to determine a winner. Assume as before that a trillion tickets have been sold. It follows that for this probabilistic setup the specificational resources include a trillion specifications and that replicational resources include a thousand possible repetitions. An elementary calculation now shows that the probability of this modified lottery having a winner is no greater than 1 in 10^{15}. That too is a tiny probability. The joint replicational and specificational resources are so inadequate that it remains exceedingly unlikely this lottery will have a winner.

In times past it used to be much easier to "inflate" probabilistic resources than it is now. The question whether the universe is finite or infinite used to be a philosophical, not an empirical question. Thomas Aquinas claimed it was only by revelation that we could know that the universe was finite. Reason, according to him, left open the possibility of an infinite universe. Spinoza's philosophical system required an infinite universe, but again on metaphysical, not empirical, grounds. Hume himself appreciated the benefits that accrue to scientific naturalism when a universe of infinite duration is presupposed:

> A finite number of particles is only susceptible of finite transpositions: and it must happen, in an eternal duration, that every possible order or position must be tried an infinite number of times. This world, therefore, with all its events, even the most minute, has before been produced and destroyed, and will again be produced and destroyed without any bounds and limitations. No one, who has a conception of the power of the infinite, in comparison of the finite, will ever scruple this determination.[12]

In his younger days Einstein had been committed to Spinoza's God. Spinoza had identified God with Nature and assumed that this God was infinite in extent and duration. Consistent with Spinoza's conception, Einstein formulated his field equations to model such an infinite universe. Now "when in 1927 the Abbé Lemaître derived from Einstein's cosmological equations the expansion of the universe and correlated that rate with data on galactic red-shifts already available,"[13] the spatio-temporal extent of the universe became an empirical question. The "data on galactic red-shifts already available" was that of Hubble and Humason. When in the early 1930s Einstein visited Hubble in California and inspected this data, Einstein came away convinced that the universe was indeed finite.[14] The inflationary universe of Alan Guth and his successors, much like the steady state theory of the 1950s, attempts to recapture Spinoza's lost infinity. In my view, these theories arise solely out of a need to preserve scientific naturalism, in this case by increasing probabilistic resources and thereby rendering appeals to chance plausible.

What event exhausts the probabilistic resources inherent in the universe? The origin of life does so quite nicely. Anyone who grapples with the improbabilities inherent in life's origin is quickly confounded. Indeed, the improbabilities are truly staggering. Fred Hoyle, for instance, computes that a single cell might on the basis of chance be expected every 10^{40000} years if the entire universe were filled with a prebiotic liquid (an assumption that is incredibly generous).[15] Bernd-Olaf Küppers, a pupil of Manfred Eigen, commenting on merely a certain subunit of a virus, writes:

> The RNA sequence that codes for the virus-specific subunit of the replicase complex consists of approximately a thousand nucleotides, . . . so that it already possess $\lambda^n = 4^{1000} \approx 10^{600}$ alternative sequences . . . . The spontaneous synthesis [of this system] . . . is therefore extremely improbable.[16]

He concludes that probability theory "does not bring us a single step further as regards the statistical aspect of the origin of life."[17] Lecomte du Noüy found similarly wild improbabilities back in the 1940s.[18] Hubert Yockey and Michael Behe continue to compute them today.[19]

Is this exhausting of probabilistic resources any reason to postulate nonmaterial intervention, to invoke a supernatural designer, or to believe in God? I have tried throughout this discussion to be cautious. My sights have ever been set on scientific naturalism. My aim has been to show that scientific naturalism is incomplete. I have sketched the beginnings of such an argument, that science cannot adequately support naturalism and that nature does things to exhaust the empirical resources determined by science. One can now try to retain naturalism by introducing a metaphysical hypothesis that postulates a lot more naturalistic stuff than science can sanction.

On the other hand, one can dispense with naturalism and introduce an entirely different sort of metaphysical hypothesis—God. These two choices do not exhaust all possibilities, but they are by far the most common.

Which is to be preferred? Since my aim has not been to pitch metaphysical hypotheses, but show that one of these metaphysical hypotheses, naturalism, cannot be redeemed in the coin of science, I shall not argue this question here. Nevertheless, it must be emphasized that science regularly has its empirical resources exhausted. Moreover, when its empirical resources are exhausted, science cannot plead momentary ignorance which it hopes some day to redress. When its empirical resources are exhausted, science is in no position to distribute promissory notes. When its empirical resources are exhausted, science itself closes the door to naturalistic explanation.

The door therefore remains wide open to a scientifically defensible account of intelligent design.[20]

NOTES

[1]Quoted in *Peter's Quotations*, s.v. "Doubt."

[2]Richard Dawkins certainly thinks so. Consider his comment on the origin of the DNA/protein machine: "[To invoke] a supernatural Designer is to explain precisely nothing, for it leaves unexplained the origin of the Designer. You have to say something like 'God was always there', and if you allow yourself that kind of lazy way out, you might as well just say 'DNA was always there', or 'Life was always there', and be done with it" [Dawkins 1987:141].

[3]Quoted in Johnson [1991:114].

[4]I owe the idea of a talking pulsar to Charles Chastain. The pulsar is an oracle. Here I am using oracles to investigate the possibility of design. Oracles, however, illuminate a host of philosophical questions. I have, for instance, used oracles to investigate the mind-body problem. See Dembski [1990:203-205].

[5]Perhaps to make this story more convincing, both the questions and the answers should be in Hebrew. I'm not sure, however, what Hebrew looks like in Morse code, so I'll stick with English.

[6]This universal bound on computational speed is based on the Planck time, currently the smallest physically meaningful unit of time. Universal time bounds for electronic computers involve clock speeds between ten and twenty magnitudes slower. See Wegener [1987:2].

[7]Even at the atomic level quantum effects make reliable storage unworkable. Indeed, the smallest scale at which vast, reliable storage is known to be possible is at the next level up—the molecular level. We can thank molecular biologists for this insight.

[8]See Balcázar [1990: chapter 11] for the underlying theory.

[9]I've chosen factoring because factoring is easy to understand. There are problems that are not just thought to be hard, but are provably hard.

[10]See Turing [1950].

[11]This example appeared first in Dembski [1991:104, note 6].

[12]Hume [1779:67].

[13]Jaki [1989:28].

[14]See Jastrow [1980].

[15]See Hoyle and Wickramasinghe [1981:1-33, 130-141], Hoyle [1982:1-65], and the appendix by Herman Eckelmann in Montgomery [1991].

[16]Küppers [1990:68]. Küppers is a pupil of Manfred Eigen.

[17]Küppers [1990:68].

[18]See chapter 3 of du Noüy [1947].

[19]See Yockey [1977] and Behe's article in this volume.

[20]Look for the upcoming book on intelligent design by William Dembski, Stephen Meyer, and Paul Nelson.

REFERENCES

Balcázar, José L., Josep Díaz, and Joaquim Gabarró.
1990   *Structural Complexity II*. Berlin: Springer-Verlag.

Dawkins, Richard.
1987   *The Blind Watchmaker*. New York: W. W. Norton.

Dembski, William A.
1990   "Converting Matter into Mind: Alchemy and Philosopher's Stone in Cognitive Science." *Perspectives on Science and Christian Faith* 42(4), 1990:202-226.
1991   "Randomness by Design." *Nous* 25(1):75-106.

du Noüy, Lecomte.
1947   *Human Destiny*. New York: Longmans, Green and Company.

Hoyle, Fred.
1982   *Cosmology and Astrophysics*. Ithaca: Cornell University Press.

Hoyle, Fred and Chandra Wickramasinghe.
1981   *Evolution from Space*. New York: Simon & Schuster.

Hubble, Edwin P.
1936   *The Realm of the Nebulae*. New Haven: Yale University Press.

Hume, David.
1779   *Dialogues Concerning Natural Religion*. Buffalo: Prometheus Books, 1989.

Jaki, Stanley L.
   1989   *God and the Cosmologists*. Washington, D.C.: Regnery
           Gateway.

Jastrow, Robert.
   1980   *God and the Astronomers*. New York: Warner Books.

Johnson, Phillip E.
   1991   *Darwin on Trial*. Downers Grove, Ill.: InterVarsity Press.

Küppers, Bernd-Olaf.
   1990   *Information and the Origin of Life*. Cambridge, Mass.: MIT
           Press.

Montogomery, John W., ed.
   1991   *Evidence for Faith: Deciding the God Question*. Dallas,
           Texas: Probe.

Peter, Laurence J.
   1977   *Peter's Quotations: Ideas for our Time*. Toronto: Bantam.

Turing, Alan M.
   1950   "Computing Machinery and Intelligence." *Mind* 59(236).

Wegener, Ingo.
   1987   *The Complexity of Boolean Functions*. Stuttgart: Wiley-
           Teubner.

Yockey, Hubert P.
   1977   "A Calculation of the Probability of Spontaneous
           Biogenesis by Information Theory." *Journal of Theoretical
           Biology* 67:377-398.

7a
Response to William A. Dembski
K. John Morrow, Jr.

IN CONJUNCTION WITH the other speakers, I wish to express my thanks to the Foundation for Thought and Ethics for putting together this symposium and bringing together people with widely differing backgrounds to exchange views on these topics. Dr. Dembski's presentation draws heavily on the physical sciences, reflecting his interests and training, which are quite different from my own.

I think it's striking that a lot of the justifications for a purposeful universe are based on analogies brought from the fields of physics and mathematics. As I mention in my own paper (chapter 10), there is an active area of consideration in science among physicists dealing with the question of the purposefulness of the universe or its design on the level of fundamental physical laws.

But my perspective on this question is quite different, and comes from my training in the biological sciences. In order to frame an appropriate response, I would like to describe some of my own background, because I think it will clarify my reaction to Dr. Dembski's presentation.

First of all, my scientific experience is in genetics and cell biology. Over the years my research has shifted more into molecular biology. My graduate training in Seattle included courses in plant evolution and population genetics, and during my post-doctoral training I was affiliated with a research institute in Italy that was largely concerned with the study of human evolution. So I look at the question of purpose in biological systems not so much from a philosophical point of view as from a point of view of practical issues; that is, for the scientists in the laboratory, how do we approach biological questions and how does the concept of evolution allow us to understand the laws that govern the living kingdom?

In agreement with Dr. Ruse, I am an enthusiastic supporter of the basic precepts of Darwinism theory. My basis for this commitment is a practical one. It has the essential qualities of a good scientific theory; it works and it provides a framework that enables us to do very good science. The practical value of the theory of evolution is one of the fundamental issues that separates people at this conference, and it deserves some further consideration.

If we look today at the events that have taken place in the past twenty years in the area of molecular biology, the progress has been nothing short of awe inspiring. To give just one example, AIDS was unknown before 1980. Now, the basis of this disease is completely understood down to the most fundamental molecular level. The virus responsible for AIDS has been mapped to the point where we know every single molecule that composes it. These advances could never have been made without a knowledge of molecular biology and, I would argue, molecular biology could not exist without the foundation of evolutionary biology on which it rests.

Most molecular biologists working in areas outside of evolutionary biology, such as biomedical science, don't think about evolution on a day-to-day level. They just do their research and are not particularly concerned with philosophical issues. Most of them are fixated on writing grants, writing papers, jumping through the scientific hoops in order to get the rewards that come with academic achievement. In framing hypotheses, developing ideas, and testing models, however, they are always working within the framework of the theory of evolution.

I will give examples from my own work. During a portion of my scientific career I have been interested in genetic variation in tumor cells. When we look at responses to anticancer drugs, we find that survival of the fittest in cultured cells follows the same rules as those used for studying fruitflies or corn plants. With some modifications we can use the same mathematical models in both systems, and can make useful predictions of which drugs will be most effective.

But many subdisciplines of biology are entirely encompassed by the paradigm of evolution. One of the interesting observations of the last few years has been the observation that DNA molecules share sequences of information even though they code for proteins with totally dissimilar functions. For example, Nobel-prize-winning work done in Dallas by Brown and Goldstein involved the receptors for low density lipoproteins. These are proteins on the surface of cells that allow entry to cholesterol. They are vital to normal fat metabolism. Brown and Goldstein studied the receptor molecule and found that a portion contains a gene sequence exactly the same as "epidermal growth factor," a growth stimulatory protein. There is absolutely no basis that anyone could have made for predicting this *a priori*. It doesn't really make sense why proteins should be constructed as patchwork from other pre-existing proteins. It makes very good sense, however, in terms of the theory of evolution if we think of natural selection as taking advantage of whatever happens to be handy at the time. In fact, the theory of evolution predicts that molecules would be made over the same way that anatomical structures are. They are taken advantage of by natural selection and made over for entirely new tasks. I might also mention parenthetically that this observation demolishes the "unlikelihood" objection to evolution. Complex structures in biology don't arise *de novo*; they evolve from pre-existing structures.

You could argue that there is no reason why a creator couldn't do that, too. Phillip Johnson has stated that God could design living creatures in any fashion he wished, including the use of natural selection. But if you accept the notion of a God who pulls out odds and ends of biological systems and throws them together (sometimes ineptly), this is a long ways from the omnipotent creator usually considered in this context. If you analyze the question through the principle of Occam's Razor, and look for the simplest hypothesis, it is easiest to envision the process of the development of living systems through natural selection. It makes sense. It works.

When we test hypotheses in biology, we are always asking, does this function or property have selective value? Does a structure or molecule increase the chances of survival for the particular organism, for the particular living system that we happen to be investigating? If it doesn't seem to, why not? Where does this lead us? To new, previously unknown functions? A cornucopia of information has appeared in the last few years in the field of molecular biology that I believe overwhelmingly supports the principles of evolution through natural selection.

Of course we could argue that a creator could design biological systems any way he wants. But if we retreat into this defense, we could equally argue that the universe began twenty minutes ago and that everything we're looking at now was put in our heads and was built around us by a wicked and capricious creator. We don't accept this idea, not because it is logically inconsistent, but simply because it has no scientific or philosophical value. It is baggage that leads us nowhere. A much simpler hypothesis is that the universe is sensible, consistent, and that all its properties can be logically interpreted.

I would assert that the whole reason for using the paradigm of evolution and natural selection is because it works for biologists, and because it has been largely responsible for the tremendous advances we've seen and the practical consequences that have accrued from developments in molecular biology. There may be molecular biologists who do not subscribe to the theory of evolution, and it may be possible for them to function creatively. But I believe that they place themselves at a tremendous disadvantage by not using the concept of evolution in the formulation of their hypotheses.

I have not presented a specific rebuttal to Dr. Dembski. His presentation is thoughtful and brings forth some salient points. But we are discussing two issues. On the one hand is the question of design and purpose in the universe. On the other is the question of the validity of biological evolution and natural selection, and its utility in a purely pragmatic sense for finding out more about the world around us and how it operates.

8
Radical Intersubjectivity:
Why Naturalism Is an Assumption Necessary for Doing Science
Frederick Grinnell

AT ONE LEVEL, Phillip Johnson's book *Darwin on Trial* [1] is a critique of modern evolutionary biology. He interprets the contradictions and controversies within evolutionary biology as evidence for the inability of scientists to understand evolution in scientific (read naturalistic or materialistic) terms, an inability he says that scientists themselves refuse to acknowledge.

Johnson's response to evolutionary biology can be understood as part of a larger tradition of religious and humanistic thought that doubts the ability of science to describe "life" according to physical and chemical ideas. For instance, until 1828 when Wohler synthesized urea, it was believed that organic matter and inorganic matter were uniquely different; that only living things could give rise to organic matter. Even after the barrier between organic and inorganic was overcome, controversy continued to surround the mechanism by which biological systems carried out complex reactions: was it enzymes or vitalistic forces? Buchner ended this dispute and won the Nobel Prize for his discovery of fermentation by cell-free yeast extracts. Afterward, doubt persisted about the chemical nature of enzymes until 1930 when Northrop crystallized the enzyme pepsin. [2] After each advance, however, the key question remained: was/is life a biochemical event, or the work of a creative intelligence?

At a more fundamental level, *Darwin on Trial* is a critique of modern science, its assumptions, its implications, and its relationship to religion. Johnson resents what he understands as the central claim of scientific naturalism:

> that scientific investigation is either the exclusive path to knowledge or at least by far the most reliable path, and that only natural or material phenomena are real. In other words, what science can't study is effectively unreal (p. 114).

He doesn't understand why science resists creationist accounts of evolution.

> In the broadest sense, a "creationist" is simply a person who believes that the world (and especially mankind) was *designed*, and exists for a *purpose*. With the issue defined that way, the question becomes: Is mainstream science opposed to the possibility that the natural world was designed by a Creator for a purpose? If so, on what basis? (p. 113).

And he thinks that the persistence of scientists in their naturalist beliefs is an attack against the importance of God and the meaningfulness of religion.

> Naturalism does not explicitly deny the mere existence of God, but it does deny that a supernatural being could in any way influence natural events, such as evolution, or communicate with natural creatures like ourselves. . . . A God who can never do anything that makes a difference, and of whom we can have no reliable knowledge, is of no importance to us (p. 115).

How should one respond to this critique of science? In his review of *Darwin on Trial*,[3] David Hull makes the following point:

> Johnson finds the commitment of scientists to totally naturalistic explanations dogmatic and close-minded, *but scientists have no choice* (my italics).

Why does Hull say that scientists have no choice?

How Scientific Discoveries Become Scientific Discoveries

There are two possibilities for understanding the absolute relationship between science and naturalism. The first possibility is to make a utilitarian argument. Science provides naturalistic explanations about the world. Modern technology, a product of science, demonstrates the "truth" of science. Therefore, naturalistic explanations are the only possible way to gain a meaningful understanding of the world. This argument fails, however. Although its practical effects indicate that science is powerful, the realization that scientific beliefs evolve over time[4] should act as an antidote to scientific hubris. Science can never have more than a limited understanding of the world.

The second possibility takes the opposite approach. It is not that science teaches us the necessity of naturalistic explanations. Rather, and here is the point to be explored in the remainder of my paper, naturalistic explanations are an assumption necessary for doing science. *Only naturalistic explanations can become part of science because of the way in which scientific discoveries become credible.*

Elsewhere I have written in detail about *The Scientific Attitude* [5] and have described three interdependent levels of action that taken together provide an account of what doing science entails. At the first level, the individual researcher works alone; at the second level, the researcher participates in scientific communities; and at the third level, the researcher lives as a person in the world. This multilevel approach is necessary to understand the cognitive features of science, the social structure of science, and the relationship between science and other aspects of human life. The researcher engages in a dialectical process whose key elements are discovery and credibility. Discovery is the first part of the dialectic; credibility is the second. *Individual scientists make discoveries; scientific communities make discoveries credible.* That is, credibility is embedded in the social structure of science.

Why must communities rather than individual investigators make discoveries credible? Remember Meno's question to Socrates: "How will you look for it, Socrates, when you do not know at all what it is?"[6] This problem perplexes every scientist. In modern terms, how do you know that what you are seeing is real? How do you distinguish between data and noise? Perhaps your results result from an artifact of your experimental design or assumptions.

Every experiment tests two hypotheses, one overt, the other hidden. The overt hypothesis is what the researcher thinks he/she is testing. The hidden hypothesis is the researcher's assumption that the experimental design, methods, and equipment are adequate for testing the overt hypothesis. One's discoveries are inextricably linked to one's expectations: expectations about what might be seen; expectations about how experiments should be done; expectations about what counts for data. Where do these expectations come from? They develop according to the scientist's education, experience, and temperament. Each investigator is unique; each investigator develops a unique style of doing research, a scientific *thought style.*[4]

No matter how convincing the results seem to be to the individual, they may be wrong. Although researchers believe that their discoveries are scientific, that their methods are reliable, and that the data are interpreted properly, they all are subject to the possibility of unrecognized error. As a result, individuals cannot verify the credibility of their own work. Rather, they must turn toward the scientific community. The credibility of my discoveries initially will depend on how convincing the results and interpretations appear to others (more on this below). And the others to whom I present my research take seriously the motto of the Royal Society, *Nullius in verba*, which P. B. Medawar translated: "Don't take anybody's word for it."[7]

Intersubjectivity

By turning toward others, scientists intuitively and implicitly move their research out of a strictly subjective framework. They transcend their subjectivity by becoming intersubjective. Intersubjectivity here refers to my recognition of others as people who are like me, whose basic experience of reality complements mine. If they were standing where I am standing, they would see something very similar to what I see. I anticipate that we share *reciprocity of perspectives*, an assumption that derives from my typical experience of the world as present not only to me but to others as well; as ours, not mine alone.[8]

Therefore, belief that observations made by one scientist could have been made by anyone makes intersubjectivity a founding assumption of the scientific enterprise. *The scientific attitude believes itself to be radically intersubjective.* Moreover, it is precisely this assumption that leads scientists to think that their observations are objective. I assume that my observations are not a result of my personal biases since I believe they can be verified, at least potentially, by everyone else. Because of this commitment to intersubjective verification, the ideal goal of science becomes inclusive knowledge. Science aims toward (although never reaches) consensus.[9]

Practically speaking, to paraphrase William James, *credibility happens to a scientific idea.*[10] Discoveries are made credible by subsequent events, events that develop in the context of the relationships between individual investigators and the scientific communities in which they participate. At first, new research is accepted depending upon how reasonable it appears. Subsequently, as the new research is used successfully it will gain in credibility. Investigators rarely replicate each others' research exactly. Rather, they use the results while pursuing their own aims. But regardless of how credible the work appears to become, the absolutely credible, or truth, remains the vanishing point in the future toward which science moves.

When research is first presented formally by an individual to the scientific community, often in the form of a manuscript submitted for publication or a grant submitted for funding, what makes the work appear credible? Reviewers of manuscripts and grants ask themselves if the results are consistent with what already is known, if the methods are contemporary, and if the procedures and findings "look" as if they could be verified. Research presented in instrumental and mathematic terms gains in credibility because it seems to depend less on personal observations. On the other hand, the more one's work appears to depend on the individual "I," the less credible it will seem. When you write a scientific paper, wrote Nobel laureate Francois Jacob, "rid [the research] of any personal scent, any human smell."[11] That is why most investigators write their papers in the anonymous third person. Since *any* investigator could have made the discovery, science denies the validity of any privileged perspectives that cannot be shared by all.

The Domain of Science

There are many different ways of experiencing the world. Here are three different perspectives about the sun. They reflect scientific, poetic, and religious attitudes.

> All the spheres revolve about the sun as their midpoint, and therefore the sun is the center of the universe.
>
> —Copernicus

> But, soft! What light through yonder window breaks? It is the east, and Juliet is the sun.
>
> —Shakespeare

> Sun, stand thou still upon Gibeon; And thou, Moon, in the valley of Aijalon.
>
> —Joshua

Why can the first statement become part of science but not the second or the third?

A well-known story tells of a night watchman who finds a man searching under a street lamp for lost keys and offers to help. Unsuccessful, the watchman finally asks the man if he is sure they are looking in the right spot. "No," came the reply, "but we can see better here."[12] *One cannot look where one cannot see. Similarly, the assumptions of science constrain those aspects of experience that can be investigated scientifically.*

How typical experiences are and how clearly they can be described together determine a continuum of what can and cannot be verified intersubjectively. Experiences that typically can be had by anyone and that can be described readily are the ones most easily incorporated into the scientific domain through the scientific attitude. Conversely, those

103

aspects of the world not subject to intersubjective verification are excluded from the scientific domain.

Consider the following Zen koan.[13]

> The wind was flapping a temple flag. Two monks were arguing about it. One said that the flag was moving; the other said that the wind was moving. Arguing back and forth they could come to no agreement. The Sixth Patriarch said "It is neither the wind nor the flag that is moving. It is your mind that is moving."

(And the thirteenth century commentator adds:)

> The wind moves, the flag moves, the mind moves: All of them missed it. Although he knows how to open his mouth, he does not see that he was caught by words.

Zen truth is lost once articulated. The Sixth Patriarch was "caught by words." The attitude of Zen experiences everyday life *directly* at a holistic level beneath and beyond any possibility of intersubjectively shared experience. Consequently, the truth of Zen is inaccessible to the scientific attitude.

For similar reasons, mystical religious experience also is inaccessible to science. The Christian philosopher and mystic Meister Eckhart wrote:

> What is contradiction? Love and suffering, white and black, these are contradictions, and as such these cannot remain in essential Being itself. . . . When the soul comes into the light of reasonableness (the true insight) it knows no contrasts. Say, Lord, when is a man in mere "understanding" (in discursive intellectual understanding)? I say to you: "When a man sees one thing separated from another." And when is a man above mere understanding? Then I can tell you: "When he sees all in all, then a man stands beyond mere understanding."[14]

At the ideal limit of the mystical domain of experience, the person becomes no-one, the world becomes no-thing, and the two fuse into an ineffable but revelatory moment.[15] In this moment, as described by Martin Buber:

> The form that confronts me I cannot experience nor describe; I can only actualize it. And yet I see it, radiant . . . far more clearly than all the clarity of the experienced world. Not as a thing among the internal things, nor as a figment of the imagination, but as what is present. Tested for it objectivity, the form is not there at all; but what can equal its presence? And it is an actual relation: it acts on me as I act on it. What then does one experience of the Thou? Nothing at all. For one does not

experience it. What, then, does one know of the Thou? Only everything, for one no longer knows particulars.[16]

For both Eckhart and Buber, true knowledge transcends the world of things, the world of contradictions, the world of logic. Encountering the presence of God means extinguishing the human intellect to the point of no-thingness. One experiences nothing by intellectual reflection, yet learns everything. Here is privileged perspective, not reciprocity of perspectives; here is an experience of "mine," not "ours." Belief is possible; intersubjective verification is not.

Conclusions

Religious faith orients the person toward the ultimate meaning of the world,[17] a meaning whose context is private and needs no inter-subjective verification for validation. How different this is from the scientific attitude that orients the person toward the possibility of an operational understanding of the world, an understanding that depends on intersubjective verification for credibility.

As much as Phillip Johnson might wish it otherwise, the sacred dimension of life witnessed by the religious attitude cannot be seen from the perspective of the scientific attitude. Because science is radically intersubjective, because science aims toward a consensus of credibility based on intersubjective verification, the naturalistic world shared by everyone is the only world accessible to science. If it can't be measured or counted or photographed, then it can't be science—even if it's important.

Acknowledgments

I am indebted to my colleagues Drs. William Snell and Richard Anderson for their advice and encouragement. Scientific research in my laboratory is supported by grants from the National Institutes of Health, CA14609 and GM31321.

REFERENCES CITED

[1]Johnson, P. E., *Darwin on Trial*, Regnery Gateway, Washington, D.C., 1991.

[2]Northrop, J. H., Biochemists, biologists, and William of Occam. *Annual Review of Biochemistry* 30:1-10, 1961.

[3]Hull, D. L., The God of the Galápagos. *Nature* 352:485-486, 1991.

[4]Fleck, L., *Genesis and Development of a Scientific Fact*, University of Chicago Press, Chicago, 1979.

[5]Grinnell, F., *The Scientific Attitude* 2nd Edition, Guilford Press, New York, 1992.

[6]Plato, *Meno*, tr. G. M. Grube, Hackett Publishers, Indianapolis, 1980.

[7]Medawar, P. B., *The Limits of Science*, Harper and Row, New York, 1984.

[8]Schutz, A., *The Phenomenology of the Social World*, tr. G. Walsh and F. Lehnert, Northwestern University Press, Evanston, 1967.

[9]Ziman, J. M., *Public Knowledge: An Essay Concerning the Social Dimension of Science*, Cambridge University Press, Cambridge, 1968.

[10]James, W., *Pragmatism and The Meaning of Truth*, Harvard University Press, 1975.

[11]Jacob, F., *The Statue Within: An Autobiography*, tr. P. Franklin, Basic Books, New York, 1989.

[12]Carmell, A., Freedom, providence, and the scientific outlook. In *Challenge*, ed. A. Carmell and C. Domb, Feldheim Publishers, New York, 1972.

[13]Shibayama, Z., *Zen Comments on the Mumonkan*, tr. S. Kudo, Harper and Row, New York, 1974.

[14]M. Eckhart quoted in Otto, R., *Mysticism East and West*, Macmillan Co., New York, 1932.

[15]James, W., *The Varieties of Religious Experience*, Macmillan, New York, 1961.

[16]Buber, M., *I and Thou*, tr. W. Kaufmann, Charles Scribner's Sons, New York, 1970.

[17]Fowler, J. W., *Stages of Faith: The Psychology of Human Development and the Quest for Meaning*, Harper and Row Publishers, San Francisco, 1981.

8a
Response to Frederick Grinnell
Peter van Inwagen

THE BODY OF PROFESSOR GRINNELL'S paper seems to me to be an argument for what I would call *methodological naturalism*. This I take to be the thesis that scientific explanations and theories should assert or presuppose the existence of nothing but natural objects. Scientific explanations, moreover, should not assert or presuppose that these natural objects have any properties but natural properties. (Some might say that a natural object like Mt. Everest has such properties as being sublime or being a divine creation, and that, unlike height and weight and other measurable qualities of things, these are not natural properties.) It may be, says the methodological naturalist, that there are objects that are not natural objects; and it may be that some natural objects have properties that are not natural properties. But such things and such properties are, if they exist, irrelevant to the enterprise of science.

Many questions might be asked about methodological naturalism. One of the most important is: What does "natural" mean? But I will simply assume that we understand this term well enough to go on.

I know, have corresponded with, and have read books by many scientists who are Christians. Every one of them is a methodological naturalist. All of them, of course, believe that there are things that are not natural things, and all of them believe that even natural things have properties that are not natural properties. Nevertheless, they would not dream of asserting or presupposing the existence of anything but natural objects and natural properties in their theories and explanations.

Methodological naturalism is, therefore, old news, and Professor Grinnell's paper is largely an argument for the truth of this piece of old news. But there is nothing wrong with that. His paper is a philosophical paper, and one of the main tasks of philosophy is to argue for old news. There are a lot of good reasons for this: arguments for old news help us better to understand our beliefs, for example, and they remind us of the value of and centrality to our thought of various beliefs that we might otherwise be as unaware of as a fish is of water.

Is the argument a good one? Well, I have heard this sort of argument before, and I have no quarrel with it. But it does strike me that there are some other things that might be said in defense of methodological naturalism.

In my own contribution to this symposium I mention a well-known episode in the history of science, the story of Newton and the instability of the solar system. I want to contrast this story with another story of more recent vintage. Several years ago, a few physicists suggested that certain effects could be explained only by the postulation of a fifth fundamental force (in addition to gravity, electromagnetism, the weak nuclear force, and the strong nuclear force). They labeled this force "hypergravity." After a short while, however, general agreement was reached that the effects the force was supposed to explain did not in fact exist, and hypergravity was removed to the scientific attic, to gather dust beside phlogiston and the luminiferous ether.

Now suppose that someone were to reason as follows. "Newton and the proponents of hypergravity each attempted to explain a certain effect by postulating something invisible to account for it—in the one case, God, and in the other, hypergravity. In each case it turned out that no account was needed, and the effort was dropped. But what is the difference between the two cases? If the postulation of a force called hypergravity (which is detectable only through the effects it is postulated to explain) is something that one can do without violating the canons of science, why is the postulation of a being called God (who is likewise detectable only through the effects he is postulated to explain) not something that one can do without violating the canons of science? What is the essential difference between the two cases? Why not, in fact, reject methodological naturalism as foundational to science, and say that scientific explanations involving God would be perfectly all right in principle—it just turns out that (as Laplace observed) they are not needed? (Not so far, at any rate. But we should recognize no fundamental objection to introducing them in the future if they should turn out to be needed.)"

I think that this reasoning is misguided, and I am not sure that an appeal to "radical intersubjectivity" does a very good job of explaining why it is misguided. To explain why it is misguided, I appeal to the following considerations.

Newton did not have a theory about God and his relation to the solar system that explained why or when or how God would correct the orbits of the planets. At any rate, he did not have a theory that explained these things in the sense that his theories of motion and gravitation explained Kepler's laws of planetary motion. According to Newton, correcting the orbits of the planets is something God "just does," and there is really nothing more to be said about the matter. The advocates of hypergravity, on the other hand, did not simply say, "There's a thing, a natural force, called 'hypergravity' and it is the cause of phenomenon X." Rather, they had a theory with a detailed mathematical structure, on the basis of which one could predict the occurrence (under conditions whose occurrence in conjunction with phenomenon X could be verified) of phenomenon X. If they *had* said, "There's a thing, a natural force, called 'hypergravity' and it is the cause of phenomenon X," and had said no more than this, then they would not have provided a scientific explanation of "phenomenon X," despite the fact that their statement appealed only to purely natural objects and properties.

The trouble with trying to construct scientific theories that appeal to God or to other supernatural agencies is, I suggest, that the "theories" always turn out not really to be theories at all. They turn out to be simple assertions, usually to the effect that some causal relation holds between God and some part of the natural world. I myself think that the statement "God is the creator of the cosmos" is true. And I think that it is a far more important truth than anything discovered by Newton, Darwin, or Einstein. But I do not mistake it for a scientific theory. It is not a scientific theory because it is not a theory of any sort. Theories tell you how things work, and this statement tells you what happened.

If the statement "God is the creator of the cosmos" is not a scientific theory, neither is the statement "Because God created it" a scientific explanation of the existence of the cosmos. It is an explanation all right, but it is not a *scientific* explanation. Scientific explanations *appeal to* theories. They are applications of theories to particular events or types of event or phenomena. The statement "Because God created it" is no more a scientific explanation of the existence of the cosmos than "Because Booth shot him" is a scientific explanation of the death of Lincoln: in neither case is a theory involved.

Thus I would supplement Professor Grinnell's argument for methodological naturalism.

It is a commonplace in discussions like this to distinguish methodological from *ontological* or *metaphysical* naturalism. Ontological or metaphysical naturalism is the thesis that everything that exists is a natural object having only natural properties. (Whatever "natural" means; remember that I have not undertaken to define this term.)

It is obvious that metaphysical naturalism entails methodological naturalism, in the sense that anyone who accepts the former is committed to the latter—one does not construct theories or explanations that appeal to things that one firmly believes not to exist. (This statement probably requires some qualification. I remember a course in colloid chemistry from my undergraduate days in which the instructor thought it permissible to appeal to "vibrations of the ether particles" in deriving some of the optical properties of colloids; this appeal was excused on the ground that the "ether particles" were, in this context, a "useful fiction.") But what are the implications of methodological naturalism for metaphysical naturalism?

I know from experience that there are people who simply *conflate* methodological and metaphysical naturalism. In a sense, these people might be said to believe that methodological naturalism entails metaphysical naturalism. But what these people are really doing is calling both theories by one name—probably "naturalism"—and are treating "naturalism" as methodological naturalism when they are called on to defend it, and as metaphysical naturalism when they are drawing conclusions from it.

Among people who are clear about the distinction between methodological and metaphysical naturalism, however, it would be hard to find anyone who thought that methodological naturalism *entailed* metaphysical naturalism. Almost everyone who is clear about the distinction between them would agree that someone could accept methodological naturalism and reject metaphysical naturalism without any logical inconsistency.

Let me offer an analogy that will help to explain why it is hard to see any logical connection between methodological and metaphysical or ontological naturalism. Professor Grinnell tells the story of a man who is looking for his keys in the light of a street lamp, even though he does not know that that is where they are. In most versions of the story, the man is a drunk, and *knows* that the keys are not in the area lighted by the lamp. That is funny. Professor Grinnell's story is not funny, however, not really, since the hero of his story is simply following the very sensible policy of not trying to use his eyes in the dark; the keys *may* be in the lighted area, and that is the only place he has any hope of finding them, so that is where he is looking. He is, one might say, an adherent of *methodological claviluminism*. But he does not accept (nor, of course, does he reject) the thesis of *ontological claviluminism*—the thesis that the keys *are in fact* somewhere in the lighted area. It is obvious that the adherent of methodological claviluminism is not logically committed to the thesis of ontological claviluminism. It should be equally obvious that the adherent of methodological naturalism is not logically committed to the thesis of ontological (metaphysical) naturalism.

Logical entailment and logical commitment are not everything, however. Some have suggested that the great and impressive mass of scientific information, explanation, and theory that are the fruit of the adherence of scientists to methodological naturalism constitutes important support for metaphysical naturalism. It has been argued that the fact that a science based on methodological naturalism has been so successful implies that the world is without "gaps" that need to be filled in by the acts of a deity: the success of a science based on methodological naturalism shows that "there is nothing left for God to do."

In my view, that argument is not cogent. In my view, it appeals to a theologically very primitive notion of what it is that God is supposed to "do." But I don't wish in these remarks to address the questions that this sort of argument raises. I will remark only that it is a philosophical argument, and that it is by that very fact highly controversial. As with any other philosophical argument, you accept it or you don't, and it is probably not going to convince anyone who is not initially sympathetic with its conclusion.

I am not sure what Professor Grinnell thinks about the relation between methodological and metaphysical naturalism. I don't see any unequivocal evidence in his paper that he thinks that his arguments (which I read as arguments for methodological naturalism) offer any support for metaphysical naturalism. There are, however, a few things that he says that make me a bit uneasy. Perhaps I have misunderstood him. I'll quote just one sentence.

> The key question remained: is life a biochemical event, or the work of a creative intelligence?

The answer I would give to this "key question" is *Yes*. That is, I think that life is *both* a biochemical event and the work of a creative intelligence. And I don't see any shadow of inconsistency or tension between these two features that I ascribe to life. I am just puzzled. I would like to know more about what lies behind the very exclusive-sounding *or* in the sentence I have quoted.

In closing, I would like to make a few comments about what Professor Grinnell says about religion. The following quotation seems to sum up his ideas. "Religious faith orients a person toward the ultimate meaning of the world." Well, yes, I can agree with that. But I think that such a statement could be very misleading. It could be taken to mean that religious faith is primarily expressed in musing on the question "What does it all mean?" or at least in some type of philosophical reflection. It suggests that religious faith consists in some sort of reaching out by the individual or the community toward a passive infinite.

My faith holds that an active Infinite is reaching out toward me and every other human being. My faith holds that there is a living reality that is an active *person*, beside which the created world (which includes at least the totality of the distribution of matter and radiation in spacetime) is, in the words of St. Anselm, "almost nothing." This active, personal, living reality has *plans* for me and for you and for everyone else, and is working to bring these plans to fruition. My faith is (so I believe) a piece of news about these plans, and it is designed (not by me; I am a mere recipient of this faith) to put me and anyone who accepts it into right relation to these plans and to their Author.

Let me sharpen these remarks about an "active Infinite" by constructing my own example of a "religious statement" about the sun. There is nothing particularly original about it; the thought behind it, if not the exact words I use, is a thought that any reasonably reflective theist would assent to. It seems to me better to reflect the religious attitude (or the theistic attitude; I am not convinced that there is any such thing as "the religious attitude," an attitude toward things that is supposedly common to, for example, Zen Buddhists and Sunni Muslims) than Joshua 10:12. That passage is a report of a speech made in the course of a narrative of Joshua's military adventures. The speech it records is not science, philosophy, or theology; it is what a novelist would call dialogue. If you wanted to compare it with something that was supposed to have come from the tongue or pen of a scientist, the famous words that Galileo never spoke about the earth (*E pur si muove*) would be a closer parallel than the words in Professor Grinnell's paper that Copernicus never wrote about the sun.[1]

But I digress. Here is my "religious statement about the sun":

> The sun exists at God's pleasure. It reflects his glory as surely as the moon reflects its light, and for that reason it is in many cultures a symbol of the divine. It exists from moment to moment only because its continued existence is his will, and it would instantly cease to exist if he stopped holding it in existence. In its interior, the principles of general relativity, quantum chromo-dynamics, and quantum electroweak-dynamics combine to produce the photons that, aeons after their production, will fall on the surface of the earth to provide the energy that living organisms will exploit. These physical laws are inventions of his, chosen freely by him, from among an unimaginable number of alternative possible sets of laws. These laws hold from moment to moment only because their continued holding is his will, and if he were to stop willing that they hold, the sun and the rest of the physical universe would instantly dissolve into chaos.

NOTE

[1]At least I don't see how he could have written them. Professor Grinnell gives no citation, and the words he attributes to Copernicus seem clearly to misrepresent Copernicus' system. His planets (since they are embedded in rotating spheres) have to move in perfectly circular orbits. At the geometrical center of each planetary orbit is a point in empty space, from which the sun (which Copernicus hardly mentions) is removed by as much as several solar diameters. The orbits of the planets, as we now know, are slightly elliptical, with their foci near the center of the sun; in consequence, a system that made the planets move in perfectly circular orbits around the sun would make wrong predictions, and they would be wrong enough to have been definitely inconsistent with sixteenth-century observational data.

9
How Incomplete Is the Fossil Record?
Leslie K. Johnson

THE CLAIM HAS BEEN made that whereas acceptable evidence of microevolution exists, there is no acceptable evidence for macroevolution. The microevolutionary changes conceded are changes in gene frequencies or genetically based adaptations, which can be demonstrated in short-term scientific studies. These include changes in the frequency of dark morphs in moths, and changes in the age of first reproduction in fish as the result of the selective actions of predators on fish.

Macroevolution, however, is seen as unsubstantiated by critics of evolutionary theory. It is not seen how a process of macroevolution could produce new higher categories of life such as bird, butterflies, and flowering plants, as well as any unique and well-developed structures they possess such as brains, wings, and flowers.

Macroevolution suffers, in this view, from unconvincing evidence, missing evidence, and counter-evidence. Deemed unconvincing is the evolutionary biologists' claim that the processes that led to observable short-term changes in the genetic complements of species (and the traits governed by these genes) also led over millions of years to bigger changes, greatly modified structures with new uses, and new kinds of organisms. Also deemed unconvincing is the occasional fossil intermediate—the odd whale with legs here and the reptile with feathers there.

The missing evidence, in this view, is explained away as gaps in the fossil record. The missing "proof" would have to be a chain from ancestor to very different descendent of adapted intermediates, not overlapping in time, each superior to its predecessor.

The counter-evidence for macroevolution is regarded to be the overlapping in time of presumed ancestral and descendent species. Other counter-evidence is held to be the apparent sudden appearance—suggesting creation—of new forms, and of life itself.

Given, finally, that the evidence for macroevolution is so bad, the reason that so many scientists stand behind it must be political. There is a struggle for cultural domination: Science or God, Evolution or

Creation. Scientists must exclude an actively creating or otherwise involved God because, if they didn't, it would mean the death of science. To win, scientists push the dogma of metaphysical naturalism, which states that knowledge can come only through the methods of investigation of natural science.

Realized Universes and Experimental Re-runs

Let us look at the first objection, that the sort of processes we observe now cannot or do not produce big changes, including novelty. Why not? Any collection of replicating elements in which new variants arise by replication error, in which different variants have different relative success rates under given conditions, will evolve. By evolve, we mean that the assemblage will exhibit change, and will produce new kinds.

An example of this is an evolutionary system set up by Professor Thomas Ray of the University of Delaware on a computer (Lewin 1992). It allows him to run the model process more than once, right from the beginning. Professor Ray programmed a digital organism that was represented by a line segment of a given length and color on the monitor screen, with a definite head and tail and a specific "genetic" code for self-replication in a sequence in between. If it could find space in the form of a physical location in computer memory, and if it could obtain an analog of sufficient energy in the form of time on the computer, it would be able to carry out the replication programmed by its code. Indeed, in a short amount of time, measured in computer generations, the screen was filled with copies of the segment.

The system was also such that random, that is, unpredictable, errors in replication would occasionally occur. These replication errors had no preordained adaptive value: they were simply random changes. After a while, segments of other lengths and slightly different sequences began to appear. This was minor novelty, so far. Then Professor Ray pressed the button to run the program overnight, and went home to bed.

What did he find the next morning? He found a high degree of diversity in his community of digital organisms. He found adaptive diversity in the form of ingenious ways of getting replicated, given the limited space and computer time and the nature of the competing segments. He found novelty, things he had not explicitly programmed that initial segment to do. There were segments that alone were unable to replicate, but in coordinated groups were able to do so (mutualistic organisms). There were organisms that consumed the code of other organisms (predators), thereby gaining time, the all-important currency in Ray's universe. There were even small segments, that through mutation had lost their own ability to replicate (they no longer had a replication code of their own), but that nevertheless persisted. They persisted inside the sequence of larger "host" digital organisms and used the host's mechanism of replication to accomplish their own (parasitic organisms). There were other small segments that actually modified the replicating of the host to serve their own parasitic needs better (virus-like organisms). There was a host that through chance errors developed a way of resisting invasion by segments that had once parasitized their forerunners. This new subtype increased at the expense of the forerunners, and came to predominate, at which time the parasites disappeared altogether, not having stumbled on a way to overcome the defense. Thus the so-called "arms race," so common in macroevolution, was underway and was observable. Overall diversity waxed, sometimes in surges of new production, and waned, sometimes in crashes of diversity.

Certainly, this is not the same as populating the planet earth with organic life. But this computer exercise succeeded in demonstrating that the evolution of complex levels of diversity is possible using only the most basic ingredients of the recipe for evolution. A simple mechanism of inheritance and replication, and a mechanism to generate variety through replication errors—when coupled with selection—were sufficient to produce a quite remarkable menagerie of digital organisms occupying very different ecological roles in the digital community. These basic ingredients were all that was needed to produce adaptive diversity quite simply and easily, as well as levels of complexity not present or anticipated at the outset.

What is nice about Professor Ray's model of evolution is that it lets us explore the question of how unique the diversity and ecological complexity of life on earth is. We have only one history of life to examine (a sample size of one). Ray's exercise is the first generation of computer experiments to understand the degree to which the patterns in evolution that occurred on earth might be more general and likely to have occurred elsewhere in the universe. Such models can also be used to test how prior evolutionary developments promote or retard the subsequent development of diversity, innovation, and complexity in an assemblage of species. In contrast, the history of life on earth is not amenable to experimentation.

So, what is there to see in the unique history of life on our planet, if we could watch the whole parade? Chains of temporally adjacent ancestors and descendents all progressively improving? Not exactly. We would see overlapping of ancestral and descendent species. Current understanding is that speciation involves the splitting of a species into two species, mother and daughter. The process is complete when individuals from the two species are no longer able to interbreed. Mother and daughter species can be contemporaries just as mothers and daughters can be. We can today find species pairs that are the two prongs of a split, like Traill's flycatcher and its sister species.

We would also not be able to show that each succeeding form was always or obviously superior to the last. This implies progression. The evidence suggests instead that quite a bit of evolutionary change is neutral, and that the latest form should not be viewed as the necessarily best in any absolute sense. Those life forms that persist into the next interval are a combination of luckier and "better than the competition" at surviving and reproducing in the given environment.

Fossils: Snap Shots from the Movie

Unfortunately, we can't watch reruns of the history of life. Instead we inspect what traces of vanished life forms we can find. How incomplete is the fossil record? How many fossil links between major groups do we expect to find? To answer this question we need to take a detour to examine why the fossil record does not and cannot give a complete phylogenetic sequence of ancestors and descendents. This is a problem in sampling theory. Let us restrict the domain of our inquiry to the Phanerozoic, representing the last 570 million years, the period during which almost all fossil organisms are found.

Paleontologists estimate that as many as 99.9% of all species that ever lived are extinct (Raup 1991). Estimates of the number of species alive today range up to 30 million, mostly insects (Erwin 1988), but more conservative estimates range from 2.5 to 5 million species. Suppose for sake of discussion we take the most conservative estimate, and state that approximately 2.5 billion total species existed over the history of the Phanerozoic.

The next point to consider is the probability of discovering or "sampling" a given species as a fossil. Many factors affect this likelihood. Organisms with hard parts are more likely to fossilize than soft-bodied ones. Creatures that inhabit aquatic or marine environments are more likely to fossilize than organisms found in dry, upland habitats. Large-bodied organisms are more likely to fossilize than small ones. Abundant or widespread species are more likely to be sampled than rare or localized species.

While these factors are important, probably the most important of all is the geological lifespan of the species. Species differ hugely in lifespans. A species that is around for tens of millions of years is much more likely to be sampled than a species lasting but a few thousand, regardless of its body size, abundance, or geographic range.

Therefore, we need to consider the distribution of lifespans of species in the fossil record. Nearly two decades ago, Van Valen (1973) constructed survivorship curves for extinct species in a number of vertebrate groups. He discovered that, to a first approximation, these curves were exponential. This implied that extinction was a stationary Poisson process, and that the probability of extinction was approximately constant per unit time. From this and later work, the average lifespan of a species was estimated to be approximately 4 million years. Note, however, that this estimate is derived from those species that were sampled by the fossil record, which is likely to be an overestimate, biased in favor of the longest surviving species to begin with. How can we correct for this bias?

The answer to this question lies in discoveries made in the subjects of community ecology and biogeography. Whenever ecologists have counted the numbers of individuals in species in ecological communities, they have found that they are best described by a lognormal distribution (Preston 1948, 1962) (Figure 1). When species frequencies are counted in doubling abundance classes (number of individuals per species in octaves), and when the sample size is large enough, a bell-shaped curve of species numbers by abundance class is observed (e.g., Hubbell and Foster 1983). Biogeographers have discovered that the ranges of plant and animal species are similarly lognormally distributed (Brown and Gibson 1983). The lognormal arises in natural autocatalytic systems such as reproducing and dispersing populations, in which many normal random factors act multiplicatively on growth. For similar ecological and biogeographical reasons, the lifespans of species are expected to be lognormally distributed as well. It is an established principle that small, local populations are more extinction-prone than large, widespread species. If we assume that risk is inversely proportional to geographic range, then lifespans would be expected to follow the lognormal. Because of the

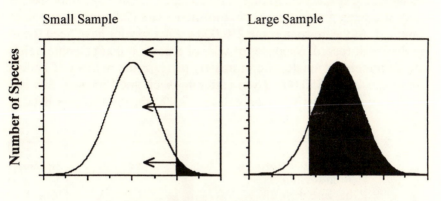

Small Sample   Large Sample

Log (2) Number of Individuals per Species

Figure 1. Lognormal distribution of the relative abundance of species. When sample sizes are small (left panel), only the very abundant species are captured (right-most tail of lognormal), to the right of the vertical line, called the "veil line" by Preston. As the sample size is increased, the veil line moves to the left, as shown in the right panel, as increasingly rarer species are added to the sample. When sample sizes are small, the visible portion of the lognormal appears approximately exponential (the righthand tail). The area of species ranges is also lognormal. Population lifespans increase with population size and geographic range, so lifespans are also expected to be lognormal.

strong impact of Van Valen's (1973) work, paleontologists have largely focused on the exponential distribution as a model of lifespan and risk of extinction, and have ignored the lognormal. However, the exponential distribution is a fair approximation to the righthand tail of the lognormal distribution. Preston (1948) noted that a nearly exponential distribution will be seen in small sample sizes of a lognormal because only the most abundant species will be sampled. Only as sample sizes increase will the mode of the lognormal be revealed. He called this effect of sample size an "unveiling" of the lognormal. It is as if a "veil line" moved from right to left across the lognormal, revealing more and more of the lognormal, ultimately unveiling the mode of the distribution (Figure 1).

The lognormal result for the fossil record is shown in Figure 2, with the placement of the veil line far toward the extreme of the righthand tail. This graph immediately reveals what a tiny sample of all 2.5 billion

Phanerozoic species the fossil record contains. On average only species with lifespans greater than 1 to 2 million years (2^{21} years) have been sampled. At the present, some 250,000 extinct species have been found in the fossil record. Suppose for sake of argument that paleontologists are extremely fortunate, and ultimately increase the number of known fossil species to 350,000. Even assuming such good fortune, this is a discovery rate of just 1 fossil species per 7,000 species that ever lived.

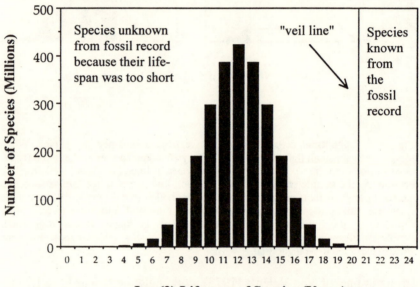

Figure 2. The lognormal applied to the data from the Phanerozoic era. Of the estimated 2.5 billion species that ever lived during this period, only extremely long-lived species were sampled by the fossil record. Because of the scaling it is difficult to see that there is still a portion of the curve to the right of the veil line. The total number of species under the curve is 2.5 billion. The number of species known from the fossil record, or ultimately knowable, is perhaps on the order of 350,000.

We can display at an expanded vertical scale just the righthand tail of the lognormal from Figure 2, to show the fact that the apparent lifespan of species in the fossil record is approximately exponential (Figure 3). When we compare the observed average lifespan of fossil species (about 4 million years) with that calculated from the righthand tail of the lognormal, we have an independent check on the time scaling of the lognormal. This gives an expected lifespan of 3.7 million years (Figure 3).

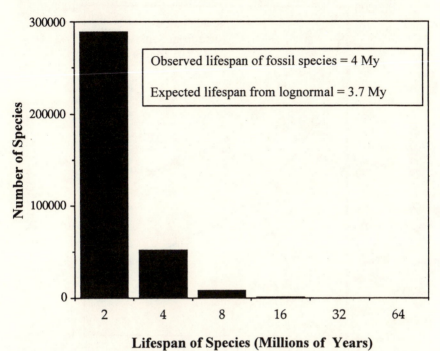

Figure 3. Righthand tail of the lognormal in Figure 2, to the right of the veil line, with an expanded vertical axis, showing the number of fossil species with given lifespans (doubling classes in millions of years). The observed and expected lifespans (4.0 and 3.7 million years) for known fossils agree reasonably well.

Although the mean lifespans are well matched, we need a more rigorous comparison of how well the expected distribution of lifespans corresponds with the data on actual lifespans of fossil species. A considerable amount of research has been done on the question of lifespans in the fossil record since the publication of Van Valen's original paper in 1973. Perhaps the best data set available is that of Professor Sepkoski of the University of Chicago, on the survivorship of 17,505 genera, as reported in Raup (1991). It would be handier to have data on individual species, but the generic data will serve our present purposes. These improved data reveal that Van Valen's assertion

121

of exponential lifespan distributions is, in fact, not precisely correct (Figure 4).

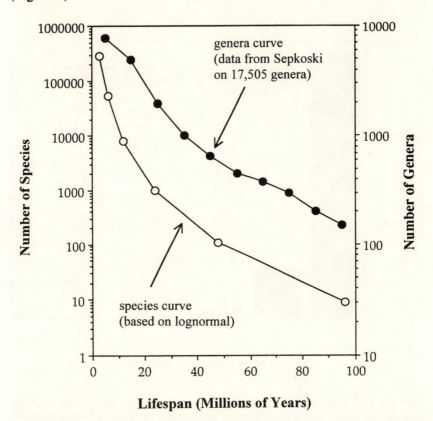

Figure 4. Comparison of the distribution of lifespan observed in 17,505 fossil genera with that expected for individual species from the lognormal. The genera curve is flatter than the species curve, as expected, since genera are expected to survive longer. Note that both curves exhibit the curvilinearity predicted from the lognormal, but not from the exponential distribution.

The actual curves of log number of surviving genera begin to flatten out when really long-lived genera are considered. This is the shape that is expected from an underlying lognormal distribution, not from an exponential distribution. The species curve predicted from the lognormal distribution is also shown in Figure 4. It shows that lifespans for individual species are shorter than for genera, as would be expected, but it also displays the curvature of the lognormal tail, not the straight line that would be expected for an exponential survivorship curve.

The main conclusion of this analysis is that fossil species are only a trivial fraction of all species that have ever lived, and therefore it is only with the greatest luck that we should find missing links, let alone a nearly continuous sequence of ancestral and descendent forms.

All this points up that the vast majority of extinct species will never be found, because they rarely met the conditions for fossilization, or because their presence was ephemeral, or because they were too rare or local in distribution. There is reason to believe, moreover, that "missing links" will be under-represented among available fossil puzzle pieces. According to current theory, the bursts of adaptive radiation that punctuate the periods of evolutionary stasis tend to involve species with more rapid generation times. Such species in turn tend to be smaller species with more delicate structures that are less likely to be preserved, and small populations at the periphery of the geographic range of the group, whose individuals are less likely to be fossilized because of their rarity.

What the paucity of links and unequivocal ancestors does not do is falsify the theory of evolution. Rather, the fossil record, for all its shortcomings, is highly supportive. What is telling is what we *don't* see: Devonian sharks with feathers and wishbones, mammals in strata with the first land plants, intermediates between trilobites and titanotheres. Instead, each new discovery corroborates the picture of the history of life in its broad outlines, a picture that makes its greatest sense in the context of adaptive evolution. A fossil whale with legs, by itself, is inadequate evidence for construction of a theory of descent by evolution. A whale with little bat wings would certainly merit concerted study, but by itself would hardly be fatal. Hundreds or thousands of such anomalies would be a serious problem, but in fact, we don't get these. What we get are new finds like the Chinese fossil bird (Sereno and Chenggang 1992), which is nicely intermediate in time and structure between *Archaeopteryx* and more modern birds. The striking thing is that all serendipitous finds fit in; the likelihood of this happening without macroevolution based on microevolutionary processes is vanishingly small.

Why Evolution Works as Science

The central point I want to make is not that the fossil record is better than it might look to some, but how wonderfully well supported the theory of evolution actually is. Let us consider what makes a theory convincing—a winner in the marketplace of ideas. The hallmark is that at heart it consists of relatively simple notions that have enormous explanatory power. The basic idea of evolution is simple: self-replication, chance variation entering the process, and differential success of the variants. Yet evolution unifies and provides a framework for nearly everything that is known in the field of biology, including most of comparative anatomy, physiology, genetics, developmental biology, much of biochemistry, molecular biology, and biogeography; and it is consistent with theories and facts in other fields, such as geology, chemistry, and physics. With evolution as a theory, it is possible to take what would be a discouragingly enormous and disparate collection of seemingly unrelated facts about organisms and to make cohesive sense of them.

A second thing lends a theory validity in the eyes of scientists. A theory gains validity to the extent it generates successful predictions in new areas, that is, in areas beyond the phenomena the theory was originally developed to explain. Another way to put it is that healthy theories pass tests coming from new directions. Validity of this kind accrues to evolutionary theory. Advances in molecular genetics, for example, have made it possible to "read" the genetic sequences of organisms. The patterns found confirm predictions of evolutionary theory.

A third feature that distinguishes theories with broad acceptance is that they point the way to profitable new avenues of inquiry. Social behavior, for example, was a long neglected field of biology. Evolutionary biologists, of course, recognized that in many circumstances individuals with genes for effective forms of parental care would have greater success in leaving descendents than individuals without such genes, meaning, therefore, that parental care could be a naturally selected trait. A breakthrough came, however, when biologist William Hamilton (1964) realized that an individual shares genes in common by descent not only with its offspring, but with all other kin as well, with more genes held in common the closer the degree of kinship. He then developed a body of ideas expressing the circumstances under which individuals would be expected to sacrifice, even to the point of ending their lives and all future chance at reproduction, for the sake of kin. William Hamilton's evolutionary thinking turned out to be seminal; it provoked the greatest burst of productive study of the social behavior of animals the world had ever seen. Indeed, kin relationships may be the single most crucial factor that structures animal societies. With evolutionary theory pointing the way, species after species was found to have the ability to recognize kin and to use this ability. Mice for example, treat experimentally introduced strangers that happen to be kin differently than they do experimentally introduced strange mice that are non-kin. They can tell kin by their smell.

I will give one last example of the productive, predictive, and unifying power of evolutionary theory. The Galápagos Islands are remote oceanic islands, never attached to any mainland. They were made, one by one, of cooled lava that welled up as a tectonic plate passed over a "hotspot," or weak place in the earth's mantle. Each island was barren at first, but was eventually colonized by terrestrial species from the mainland or from neighboring islands. Descendents of the colonists then evolved *in situ*.

There are two lineages of iguana on the islands, the marine iguana and the land iguana. Whereas no presently existing Galápagos island is more than three million years old, immunological studies on the iguanas by Wyles and Sarich (1983) indicate that the evolutionary divergence of the lineages must have occurred 15-20 million years ago. No source population of these iguanas on the South American mainland is evident.

Instead of jettisoning evolutionary theory, the investigators took a new tack. They suggested that older islands in the Galápagos chain may have existed, but became submerged. Sure enough, Christie *et al.* (1992) report finding drowned islands downstream from the Galápagos hotspot. These islands extend the time for speciation another 2-6 million years. Moreover, on the basis of geophysical evidence, the geologists predict that older islands yet will be found. They consider it likely that Galápagos islands have existed during the entire 80-90 million year history of the hotspot.

This example shows all the kinds of strengths exhibited by the theory of evolution: consistency with information in several fields (immunology, biogeography, and geophysics), predictive power, and independent corroboration. To scientists, the theory of evolution is convincing on its own merits. Nothing else needs to be invoked, including desperation or adherence to dogma.

Scientists are not disciples of metaphysical naturalism, which holds that science is the only way of knowing. All scientists have other ways of knowing if only because they are people with human thoughts and

125

concerns. In addition, scientists span the range of religiousness, from the truly devout for whom the everyday world is suffused with spiritual light, to active atheists.

Phillip Johnson in the 1990 booklet *Evolution as Dogma* (Dallas: Haughton), is concerned by the notion that science is absolute truth, which might imply that other forms of knowledge are fantasy. This is a misreading. Science tells us what the natural world appears to be like, based upon our senses and the instruments we devise to extend our senses, as probed using scientific methods in our tiny corner of space-time. The scientific method leads us to ask questions and to test hypotheses, and few would question its practical and aesthetic contribution to human welfare. Anyone who says that scientific knowledge equals Absolute Truth, however, is confusing the map with the territory.

Those concerned that science leaves no room for God are similarly misled. On a number of profound issues and concerns, science is of no use at all. On the compelling question of how to be a good human being, science is silent. On the question of the meaning of existence, it is likewise mute. Nonscientific ways of knowing are crucial to our well-being. Nevertheless, I myself am flattered (a human response) to think that we humans are capable of creating for ourselves a complex reality that includes science, and I would hope that the complex reality of other people can also include science. Then, our moral sense—separate from science—can guide our use of scientific knowledge for the welfare of humankind and our planetary home.

Past, Future, and Present: Conservation Is the Link

Regardless of our different views about origins, we need to be concerned about what is happening to life. We should be especially concerned about slowing the ongoing mass extinction of species. Estimates suggest that tropical deforestation will result in the extinction of perhaps one-fifth of all species on earth within the next one hundred years. If this comes to pass, this rate of extinction will have been more than a million times faster than the average rate of extinction during the Phanerozoic.

Faced with the enormity of the present extinction crisis, the question of whether species were specially created or evolved seems almost quaint. What better goal to bring the communities of creationists and evolutionary biologists together than to commit to saving the wonderful diversity of life on earth? Creationists should be among the most ardent conservationists of them all.

LITERATURE CITED

Brown, J. H. and A. C. Gibson 1983. *Biogeography*. Mosby, St. Louis.

Christie, D. M., R. A. Duncan, A. R. McBirney, M. A. Richards, W. M. White, K. S. Harpp and C. G. Fox 1992. Drowned islands downstream from the Galápagos hotspot imply extended speciation times. *Nature* 355:246-248.

Erwin, T. L. 1988. The tropical forest canopy. In: E. O. Wilson (ed.) *Biodiversity*. National Academy Press, Washington, D.C.

Hamilton, W. D. 1964. The genetical evolution of social behaviour. *J. Theor. Biology* 7:1-32.

Hubbell, S. P. and R. B. Foster 1983. Diversity of canopy trees in a neotropical forest and implications for conservation. In: S. L. Sutton, T. C. Whitmore and A. C. Chadwick (eds.) *Tropical Rain Forest: Ecology and Management*. Blackwell, Oxford.

Johnson, P. 1990. *Evolution as Dogma: The Establishment of Naturalism*. Haughton Publishing Co., Dallas.

Lewin, R. 1992. Life and death in a digital world. *New Scientist*: 36-39.

Preston, F. W. 1948. The commonness, and rarity, of species. *Ecology*: 29:254-283.

Preston, F. W. 1962. The canonical distribution of commonness and rarity. *Ecology* 43:185-215.

Raup, D. M. 1991. *Extinction: Bad Genes or Bad Luck?* W. W. Norton, New York.

Sereno, P. C. and R. Chenggang 1992. Early evolution of avian flight and perching: New evidence from the Lower Cretaceous of China. *Science* 255:845-848.

Van Valen, L. 1973. A new evolutionary law. *Evolutionary Theory* 1:1-30.

Wyles, J. S. and V. M. Sarich 1983. Are the Galápagos iguanas older than the Galápagos? Molecular evolution and colonization models for the archipelago. In: *Galápagos Organisms*. Pacific Division of the AAAS, San Francisco.

9a
Response to Leslie K. Johnson
Evolution as History and the History of Evolution
David L. Wilcox

MY PRIMARY CRITICISM of Leslie Johnson's interesting paper will, I'm sure, be unexpected, even thought impolite. Her primary focus is irrelevant to the topic of this symposium. But that irrelevance is highly significant. It is clear that I need to justify such an outrageous statement. Dr. Johnson makes an eloquent plea for the adequacy of the fossil record in documenting descent, and that is just the problem. Common descent is not particularly relevant to our theme, the reason for the acceptance of neo-Darwinism. "Evolution" and "Darwinism" are not synonyms, and Darwinism is not a theory *that* creatures share common ancestors—even structurally very different ancestors—it is rather a theory of *how* those creatures became different. Darwinism is a theory of mechanism, not a proposed historical scenario.

I realize that words like *evolution* and *Darwinism* have been used like the walnut shells in a shell game to obscure distinctions and to persuade the faint in heart. *Evolution* has at least been used to mean change, mechanism, history, paradigm, and world view, more linguistic freight than any word can meaningfully carry. However, neo-Darwinism—the Modern Synthesis—is *supposed* to be a theory of a mechanism by which genotypes (and the phenotypes they produce) can be transformed. Evidence that links two different forms of life in common descent simply describes a phenomenon that we need to explain.

This probably seems like nit-picking to a 1990s audience, but it would have made perfect sense to one in the 1890s. The Modern Synthesis has been so generally accepted that it has practically become synonymous with all the various meanings carried by the word *evolution*. To evaluate *why* this confusion exists, we must disentangle those meanings.

A quick trip to pre-Darwinian England can help to clarify the confusion. One hundred and fifty years ago, according to Gillespie

128

(1979), most naturalists accepted the idea of common ancestry, but they differed on how new forms arose. The Establishment at Oxford (Buckland, for instance) evidently thought that God occasionally remodeled an existing form into a perfectly adapted new type (Rupke, 1983). The Radical Materialists such as Grant and Knox followed Lamarck in considering matter itself energized with an intrinsic tendency for uniform development (Desmond, 1989). The followers of German *Naturphilosophie* (Richard Owen, for instance) held the theory that autonomous extra-material archetypes shaped lineages progressively into their own images (Desmond, 1982). All the schools (with the exception of Louis Agassis) viewed fossil sequences as demonstrations of common descent. They differed on the nature of the power that shaped biological form, but *not* on whether things shared common ancestry. One further note: although they differed in their philosophies of nature, each school had both Christian and non-Christian adherents.

According to historian James Moore (1982), however, around 1840 a new movement of young middle-class reformers calling themselves "Naturalists" appeared. This group as young adults typically changed their creed from Christianity (which they felt was morally bankrupt) to one based on "Nature." They were "poets and lawyers, doctors and manufacturers, novelists and naturalists, engineers and politicians." The group included such well-known individuals as George Eliot, Herbert Spencer, Matthew Arnold, Francis Galton, J. A. Froude, G. H. Lewes, Charles Bray, Alfred Lord Tennyson, John Tyndall, F. W. Newman, A. H. Clough, Harriet Martineau, F. P. Cobbe, and, of course, T. H. Huxley. Moore shows that the central feature of this new creed was the redefinition of human nature, society, order, law, evil, progress, purpose, authority, and nature itself in terms of the Naturalists' particular view of Nature, as opposed to the Christian Scriptures. In fact, they tended to attack the Christian Scriptures as the true source of societal evil. God, if he existed, was to be known only through the Nature which he made. Thus, according to Moore (1982) and Young (1980), "positivism" was not primarily a methodology for science, but a religious movement that sought to replace the cultural dominance of the Established Church.

Charles Darwin launched his theory of biological change in this context. He proposed a mechanism for the appearance of new forms that did not depend on any pre-existing or exterior shaping forces. The environment became the only needed constraint. It was a theory of strategic importance for the Naturalists, particularly for the "X" club, Huxley's "Young Guard" party in science.

The significance of a mechanism can be understood only within the world views of its proponents. The "Naturalism" that initially proposed and supported Darwin's mechanism was both a world view and a social movement. These individuals viewed the world as autonomous, and the Darwinian mechanism as autonomous creator. The scientific members of this movement, Huxley's "X" club, were engaged in a successful campaign to wrest the university chairs in the sciences from the clergymen/naturalists of the Established Church. The ability of Darwinism to replace the divine with a natural process was a critical support.

Turner (1978) has proposed that this fabled Victorian conflict was primarily a "professional" struggle for scientific autonomy and authority, a struggle between the "professionally" trained and validated scientists and the Anglican dons. Still, if the professionally validated "scientist" is viewed as the only one who can adequately understand nature, and if *Nature has replaced Scripture* as the source of moral and teleological truth, *ipso facto* the scientist has replaced the priest. Thus, the "professional" position at stake was as much the pulpit as the lectern.

Thus, although in reality it is just a simple proposal of natural processes, Darwinism historically *was* accepted by the Naturalists by philosophical preference. Huxley himself did not accept its scientific inference for the fossil record until after 1864 (Desmond, 1982). Indeed, as a "scientific inference," a description of material cause, other schools of thought also accepted Darwin's mechanism, but they considered it inadequate as an explanation of important biological change. Neo-Lamarckians such as Cope, and Mutationalists like De Vries, held competing theories of mechanism for morphogenesis.

In particular, Christian theists who held the universe to be governed at all points, rather than autonomous at all points, simply took the mechanism to be an aspect of God at work (Livingstone, 1989). This view I want to highlight for a moment, since it directly bears on the "blind watchmaker" question. Such men included the "American Darwin," Harvard botanist Asa Gray, who introduced and defended Darwin's theory to America, and the conservative Princeton theologian, B. B. Warfield. But Gray said, "If Mr. Darwin believes that the events which he supposes to have occurred . . . were undirected and undesigned . . . no argument is needed to show that such a belief is atheistic." Warfield (1988) commented:

> Mr. Darwin's difficulty arises on one side from his inability to conceive of God as immanent in the universe and his consequent total misapprehension of the nature of divine providence, and on the other from a very crude notion of final cause which posits a single extrinsic end as the sole purpose of the Creator. No one would hold to a doctrine of divine "interpositions" such as appears to him here as the only alternative to divine absence. And no one would hold to a teleology of the raw sort which he has here in mind—a teleology which finds the end for which a thing exists in the misuse or abuse of it by an outside selecting agent.

Even Charles Hodge, a theologian who attacked Darwin, did so because he said Darwin intended by the term "natural" selection to exclude "supernatural" selection. According to Hodge (1874),

> It is however neither evolution nor natural selection which give Darwinism its peculiar character and importance. It is that Darwin rejects all teleology, or the doctrine of final causes. He denies design in any of the organisms in the vegetable or animal world."

Hodge rejected not the mechanism, but the theological hypothesis of the blind watchmaker.

Darwin did not publish his rejection of the design argument until 1868 at the end of *Animal and Plants under Domestication*. Using the analogy of a building constructed from the stone fragments at the base of a precipice, Darwin stated:

> In regard to the use to which the fragments may be put, their shape may be strictly said to be accidental . . . Can it be reasonably maintained that the Creator intentionally ordered, if we use the words in any ordinary sense, that certain fragments of rock should assume certain shapes so that the builder might erect his edifice? . . . we can hardly follow Professor Asa Gray in his belief "that variation has been led along certain beneficial lines" . . . On the other hand, an omnipotent and omniscient Creator ordains everything and foresees everything. Thus we are brought fact to face with a difficulty as insoluble as is that of free will and predestination.

According to Gray's school of thought, the Darwinian mechanism could be used to *support* the existence of God. But can you imagine any scientist saying that in public today?

The Naturalists succeeded. The "Young Guard" used the trappings of religion to sacralize their "science." Three centuries of cooperation between science and religion were forgotten and their history was rewritten as "warfare." Hymns to nature were sung at popular lectures before the giving of "lay sermons" by a member of Galton's "Scientific Priesthood." Museums were built to resemble cathedrals, and following frantic string-pulling by Lubbock (a member of the "X" club) Charles Darwin was buried in Westminster Abbey. The new church was established (Moore, 1982).

In her paper Dr. Johnson objects that many scientists are religious, which is of course true. But, the ongoing success of "scientific naturalism" as a religious movement can be judged by the present general acceptance by "Science" and by the "Public" of the pronouncements of those "true believers" of the "church scientific" who still exist and evangelize among us. E. O. Wilson (1978) is clearly acting in a clerical role when he tells us:

> This mythopoeic drive [i.e., the tendency toward religious belief] can be harnessed to learning and the rational search for human progress if we finally concede that scientific materialism is itself a mythology defined in the noble sense . . . Make no mistake about the power of scientific materialism. It presents the human mind with an alternate mythology that until now has always, point for point in zones of conflict, defeated traditional religion . . . The final decisive edge enjoyed by scientific naturalism will come from its capacity to explain traditional religion, its chief competitor, as a wholly material phenomena.

The societal clout and ability of scientific naturalism to marginalize its competitors has been evaluated by sociologist of science Eileen Barker (1978), who concludes:

> The Biblical literalist, the Evangelical revivalist, the political visionary and even the slightly perturbed old priesthood of the established theologies turn to the new priesthood [of science] for reassurances that their beliefs have not been left behind in the wake of the revolutionary revelations of science. The new priesthood has not been found wanting. Sometimes with formulae, sometimes with rhetoric, but always with science, the reassurance is dispensed.

Again, can you imagine any scientist saying in public today that the Darwinian mechanism supports the existence of God? Don't misunderstand me. I am not suggesting they should. I am sure you will agree that scientists should leave the mention of God out of their writing, and just discuss science. However, until, for instance, the AAAS comes out with a public statement censuring such mention in the writing of popular spokesmen for science, it remains a critical issue. It is a fact that God is continuously being publicly discussed by very well-known scientists—just read Gould, Dawkins, Hull, Provine, Wilson, Simpson, Futyama, Sagan, Hawking, and others. From a nineteenth century perspective, books like *The Blind Watchmaker* (Dawkins, 1986) and *Wonderful Life* (Gould, 1989) are simply Bridgewater treatises such as Paley, Owens, and Roget wrote, works in which up-to-date science is used for the task of world-view apologetics.

In such a climate, it is a trifle hard to be objective with the data—which are all viewed as support for the dominant paradigm/world view; hence, Dr. Johnson's use of the evidence for descent.

But what do we need from the fossil record if we are to test for the adequacy of Darwinism (defined as mechanism)? Neither proof for descent nor for transformation; they might have other causes. Rather, we need evidence for the action of the environment in selecting that form, evidence that the environment has acted as a pattern-fitting mechanism—that is, evidence of the causes that *produce* morphological change. And if we want to test the blind watchmaker world view, we need evidence that demonstrates that such changes are unguided. We must explain the cause and pattern of the appearance of biological novelty. In that light, I have a few other comments or questions.

First, a minor point. It is true that new fossils fit the patterns predicted by the evolutionary sequence (but that's not particularly relevant). That pattern, however, was proposed long before the general acceptance of descent with transformation, not to speak of Darwin's theory. Those who proposed it were clearly working from *some* sort of hypothesis other than complete randomness. No one has ever thought *that*, and using it to "test" for evolution is testing against a straw-man. How can you know where groups would be placed if they arose independently? Why suppose random placement? Would not an intelligence be the *expected* source of new groups in such a case? Why would an intelligence use a pseudo-random scatter of appearances? The fact that such groups fit into the accepted patterns is proof only for some sort of shaping pattern—it is not even proof of descent.

Second, the major support adduced by Dr. Johnson for the neo-Darwinian hypothesis is the model world created by Thomas Ray (1991) of the University of Delaware. As an old "model builder," I

133

would love to get my hands on Ray's intriguing model. Nevertheless, I don't think that it is an adequate proof of the Darwinian hypothesis. Models never are. Rather, it explores the implications of its instruction set—and *that* is the equivalent of "raw" fitness values, reproductive information with no tie to a phenotype. The world of Ray's critters, the computer itself, must be programmed *into* the instruction set for it to be a real equivalent. Ray's computer is more than a coherent and limited environment. With "energy" gaining and "replicative" machinery built into the computer, and with those fundamental aspects of the model unable to be mutated, the computer is the equivalent of an infected cell, an electronic host in which viruses live. Maynard Smith (1992) considers it the equivalent of an "RNA" world, with no distinction between phenotype and genotype. But, an RNAzyme has both: nucleotide sequence (genotype) and molecular surface (phenotype). In Ray's world, the reproductive "phenotype" is built into the computer, and the "virus" just gives it instructions.

Also, Ray's critters produce no encoded morphology. The various critters produced vary only in their particular variant of the programmed commands for "reproduction." Although the outcomes are intriguing, all the complexity produced is "economic" rather than "morphological." The model *does* suggest that parasitism is a logical and necessary implication of a world with reproduction, rather than an ethical issue. But, the model is not truly open-ended. If it was, Ray would not have to be planning to add new instructions for sex and multicellularity. The program would write them for itself. It would produce its own Cambrian explosion (Lewin, 1992).

Two final notes: to avoid being swamped by inviable changes, and thus to allow mutants that could survive, Ray specifically limited his instruction-set to 32 possible mutant changes from a possible instruction-set of 10^{11}. But even that full instruction set is equivalent to the probability space of only 37 DNA bases. Thus, this random walk occurs in an unrealistically limited probability space.

In addition, the genomes of flesh-and-plasma organisms contain cybernetically error-checked programs for the production of morphologies. Thus, real genomes constrain encoded instructions of at least two classes, prescriptive and adaptive. Ray's critters have neither. Fascinating they are, significant in some ways, but they are no particular proof of the ability of neo-Darwinian mechanisms to produce novel structures.

Finally, Dr. Johnson defends evolution in terms of its importance for biology, pointing to its unifying, predictive, and productive capabilities. That statement raises intriguing questions about the nature of science. Is it true that biology cannot live without evolution, that (to quote Dobzhansky) "nothing in biology makes sense except in the light of evolution"?

First, *is* it the unifying theory in modern biology? Why not cell theory? or molecular energetics? or hierarchy theory? or ecosystem dynamics? or cybernetic control theory? The fact that a theory *applies* to all living things does not mean that it is the essential *organizing framework*. In reality, what is probably meant by evolution in this "unifying" context is simply philosophical materialism, but that is general philosophy rather than science.

Second, *is* it all that predictive? It is true enough that Sereno and Chenggang's (1992) new birds fall into the right gap, but that is not an effective logic for rejecting a theory that makes the same predictions. For instance, Richard Owen's nineteenth century theory of metaphysical archetypes would have "predicted" the same findings (or at least their probability). As I have already pointed out, the prediction of a designed universe is not the appearance of new morphologies in a random scatter.

Third, *is* the productivity of the theory evidence for its validity? The evidence of history is that *any* new and widely accepted paradigm leads to a furious round of research and scientific advance. It was, after all, the "higher anatomy" of the Idealists that led to the science of comparative anatomy. Certainly it is not the productivity of the theory of the spontaneous generation of life that has kept that field so busy.

In reality, we all do science caught between our world views and the hard-edged facts of the real world. But that tension is buffered by a hierarchy of progressively more inclusive theoretical lenses through which we view the world. We investigate the real world under the guidance of our recognition frameworks. As Stephen Gould put it (1980):

> First, facts do not come to us as objective items seen in the same unambiguous way by all reasonable people. Theory, habit, prejudice and culture all influence the facts we choose to observe and the way in which we perceive them. Second, the construction of theories is not a "second story" operation in science, an activity to be pursued after constructing a factual ground floor. Theory informs any good scientific work from the very beginning; for we ask questions in its light, and science is

inquiry, not mindless collection. Moreover, the sources of theory are manifold; new ideas arise more often by the creative juxtaposition of concepts from other disciplines . . . than from the gathering of new information within an accepted framework.

REFERENCES

Barker, E. "Thus Spoke the Scientist: A Comparative Account of the New Priesthood and its Organizational Cases." *Annual Review of the Social Sciences of Religion* 3 (1979): 99.

Dawkins, R. *The Blind Watchmaker.* New York: W. W. Norton and Co., 1986.

Desmond, A. *Archetypes and Ancestors.* Chicago: University of Chicago Press, 1982.

Desmond, A. *The Politics of Evolution.* Chicago: University of Chicago Press, 1989.

Gillespie, N. C. *Charles Darwin and the Problem of Creation.* Chicago: University of Chicago Press, 1979.

Gould, S. J. "The Promise of Paleobiology as a Nomothetic, Evolutionary Discipline." *Paleobiology* 6.1 (1980):96-118.

Gould, S. J. *Wonderful Life.* New York: W. W. Norton and Co., 1989.

Lewin, R. "Life and death in a digital world." *New Scientist* 22 (1992): 36-39.

Livingstone, D. N. *Darwin's Forgotten Defenders.* Edinburgh: Scottish Academic Press, 1987.

Moore, J. R. "1859 and all that: Remaking the Story of Evolution-and-Religion." *Charles Darwin, 1809-1882: Centennial Commemorative.* Wellington, N.Z.: Nova Pacifica, (1982): 167-194.

Ray, T. S. in "Artificial Life II." Santa Fe Inst. *Studies in the Sciences of Complexity.* Vol. XI. Eds. Langton, C. G., S. Taylor, J. D. Farmer and S. Rasmussen. Redwood City, CA: Addison-Wesley, (1991): 371-408.

Rupke, N. A. *The Great Chain of History.* Oxford: Clarendon Press, 1983.

Turner, F. M. "The Victorian Conflict Between Science and Religion: A Professional Dimension." *Isis* 69 (1978): 356-376.

Warfield, B. B. "Charles Darwin's Religious Life." *Princeton Theological Review* (1888): 569-601.

Wilson, E. O. *On Human Nature*. Cambridge, MA: Harvard University Press, 1978.

Young, R. M. "Natural Theology, Victorian Periodicals and the Fragmentation of a Common Context." *Darwin to Einstein: Historical Studies on Science and Belief*. Editors. C. Chant and J. Fauvel. Harlow: Longman, 1980. 64-107.

Sereno, P. C. and R. Chenggang. "Early Evolution of Avian Flight and Perching: New Evidence from the Lower Cretaceous of China." *Science* 255 (1992): 845-848.

Smith, M. "Byte-sized evolution." *Nature* 355 (1992): 772-773.

10
Teleological Principles in Biology:
The Lesson of Immunology
K. John Morrow, Jr.

I AM PLEASED TO have an opportunity to take part in a lively and broad-ranging symposium dealing with the central theoretical principle of biology. I come here with much interest and anticipation, for if I could establish that the theory of evolution is invalid, it would be the greatest scientific discovery of the twentieth century. I would go down in history as one of the greatest *savants* of all time. A new era, a new age of science would be ushered in. Inconceivable wealth and power would be mine. If I could be even peripherally associated with such a monumental event, it would assure my future.

Theodosius Dobzhansky said in 1973, "Nothing in biology makes sense except in the light of evolution."[1] That assessment, coming from one of the leading biologists of the twentieth century, asserts that all of biological science collapses into a jumble of unrelated measurements and observations if not linked together by the central guiding generalization of evolution.

But biological science, chemistry, physics, and other branches of modern scientific inquiry are also joined together by a common purpose, rationale, and experimental approach. Thus a rejection of evolutionary theory would pull the rug out from under the whole edifice of modern science, allowing it to crash down in rubble all around us.

Because of all that hangs in the balance, it behooves us carefully to evaluate and question any attack on the theory of evolution, and to subject it to the most thorough scrutiny.

Professor Phillip Johnson's frontal assault on evolutionary theory, because of the substantial attention that it has received, deserves a careful critique, and if found unsatisfactory, deserves a measured rebuttal.

There is another reason, however, for a rigorous evaluation of evolutionary theory at this time. Molecular biology has made tremendous strides in the last ten years, and these techniques and approaches have been applied to the design of evolutionary trees, analysis of relationships between taxonomic groups, measurements of chronological separation of species, comparison of genomes in different taxonomic groups, and to a variety of issues in evolution. Workers in the field argue that this information imparts support and a new understanding of evolutionary theory, whereas Professor Johnson asserts that it merely shows similarities and has no bearing on the validity of evolutionary theory. In light of such conflicting views it is even more appropriate to reexamine the underpinnings of evolutionary theory at this time.

I do not wish to participate in a Johnson bashing, or a microanalysis of the arguments presented in his book. I do not believe that arguments over technical points in paleontology, molecular biology, or ecology will resolve the basic issue. Rather I wish to consider on a fundamental level the paramount philosophical question: the concept of purpose in biology.

The idea that the universe is ruled by a purposeful, guiding hand has been with us for a long time. Clearly this represents the primary issue with which we are concerned. Either mechanistic explanations are sufficient to explain the diversity, the function, and the beauty of the living world around us, or they are not.

But it is difficult to focus on the issue of purpose as it relates to a concept as broad and diverse as the study of evolutionary biology. For this reason I choose to examine the role of teleology in a much more restricted segment of biological science, which I feel qualified to discuss in detail and which is thoroughly understood on a molecular level, the field of immunology. I wish to focus particularly on the historical development of our understanding of how antibody molecules are generated. In doing so, I believe we can judge the current arguments concerning the adequacy of mechanistic evolutionary theory to explain the diversity of living forms.

Immunology, like other areas of biology, has developed rapidly in the last twenty years and has profited immensely from research using the techniques of molecular biology. In doing so, it has changed from a discipline that was mainly descriptive and that lacked clear models to a stage in which precise cause and effect relationships of the most minute events occurring within the immune system can be understood.

In the 1950s none of this was possible. It was known (and had been known for many years) that when humans or animals are exposed to foreign substances such as large protein or carbohydrate molecules (usually through introduction into the circulation) they will produce a protective response in the form of circulating proteins. The foreign substance is known as an antigen, and the protein produced by the organism in response to it is known as an antibody. Antibodies are large molecules that constitute one of the main lines of defense against marauders from outside. Any question of their value for survival is answered by the pathology of AIDS in which the immune system is despoiled by HIV.

Moreover, these antibodies were known to combine with their respective inciting antigens in a reaction of great specificity. A lock and key mechanism was proposed by which the antibody molecule came precisely to fit the shape of the antigen. Following this binding it could, with the aid of other reactions, eliminate the offending antigen from the host.

Antibodies possessed an almost mystical quality. They were available in practically endless variety. For instance, Landsteiner had shown early in the century that an immune serum could distinguish between two proteins having as little as a single amino acid difference between them. Even finer levels of recognition were possible between D and L amino acids and ortho and para positions on benzene rings. This meant that the host must have the capacity to produce a virtually unlimited repertoire of responses. If one considers all the possible ways in which simple molecules can be modified and that the host should be able to generate antibodies to each of these modifications, then clearly there must be literally millions, and perhaps billions, of possible antibody types.

Three major theories were developed to explain how the tremendous diversity of the immune system could be generated. These three hypotheses were the instructive (or template) theory, the subcellular selection (or somatic mutation) theory, and the germ line theory. Elements of the latter two theories were later combined by Burnet in his "Clonal Selection Theory."

The instructive or template theory was proposed in its modern form in about 1930 by Felix Haurowitz and Linus Pauling (Mazunder, 1989; Kindt and Capra, 1984). It proposed that the antibodies were molecules that behaved in a plastic, flexible manner, so as to mold themselves to the shape of the antigen. Pauling, in 1940, suggested that the final step of synthesis of the antibody would be to fit the antigen. Thus the antibody was envisioned as an all-purpose, amorphous blob that after embracing the antigen was frozen into the mirror image of the antigen.

The template hypothesis was devised to get around the problem of storage of vast quantities of information. It got into trouble immediately, because it did not explain one of the most significant phenomena of immunology, that of vaccination. It had been recognized for hundreds of years that when an individual is exposed to a pathogenic agent such as smallpox virus, after a bout of disease the individual is immune to subsequent attacks of the same disease. Lady Mary Montagu, wife of the British ambassador to Constantinople in the early 1700s, is usually credited with bringing the discovery to England. She did this against the recommendation of her clergyman, who felt that vaccination against smallpox would be effective only in the heathen. In the nineteenth century Pasteur and others put the principle on a firm footing.

Thus if the instructional hypothesis were correct, it had to explain the fact that the immune system possessed a memory that could last sometimes for decades.

Its failure to explain immune memory satisfactorily cast the template theory in doubt from its inception. It was dealt a fatal blow by the new findings in the field of protein chemistry, which established that the folding of a protein molecule was a consequence of its amino acid content, which in turn was specified by its genetic program. Since the antibody molecule was composed of amino acids coded by its particular genetic message, there was no way in which it could change its shape, which was permanently fixed at its time of synthesis.

The second hypothesis was the somatic mutation hypothesis put forth in its modern form by Joshua Lederberg (1959) and others. This term refers to genetic modifications occurring in somatic or body cells, and not passed on through the germ line, i.e., the union of egg and sperm that binds each generation to the next. This hypothesis also possessed major drawbacks. It assumed that only a limited number of genes specified the structure of antibody molecules. These genes, however, were highly unstable and could mutate through an incredible variety of types, generating all possible antibody conformations. Thus in each generation the host would generate an entire panoply of antibodies, and the antigen would select and cause the amplification of the cell carrying its complementary antibody.

Although this theory escaped the problem of having to propose an antibody that could wrap around every conceivable antigen, it required two ad hoc assumptions: first, that antibody-forming cells were capable of exceedingly high mutation rates, and second, that the antigen could somehow select one cell, programmed to form the appropriate antibody from a great mass of cells producing irrelevant antibodies. Presumably following this selection there would be a stimulation of the relevant cell to generate a large population of descendants. At the time both of these problems were substantial roadblocks to an acceptance of the theory.

The third theory, known as the germ line theory (Kindt and Capra, 1984), proposed that all the information for producing all antibody types was carried in the genes, and passed from generation to generation through the germ line. Under the most direct form of this hypothesis, there would exist a separate gene for the antibody that reacts against every single molecule in the known universe. As I have suggested above, this would require an immense amount of genetic information. Some workers suggested at the time that this could account for the fact that mammals have so much more DNA than bacteria; i.e., aside from the housekeeping functions that they share with the bacteria, almost all their resources are given over to coding for antibodies.

The germ line theory also had a number of problems, not the least of which was the immense amount of baggage that an organism would be saddled with, and which, in the vast majority of instances, would never be called forth. Further, the theory was imbued with a strong element of teleology. It implied that mice, rabbits, and humans "knew" in some molecular sense that synthetic compounds would be invented millions of years before those compounds existed. By pursuing this line of reasoning, one would be forced to conclude that, even now, organisms carry with them the genetic blueprint for molecules that have yet to be synthesized in the laboratory of some as-yet-unborn organic chemist.

Between 1957 and 1959 F. MacFarlane Burnet proposed the clonal selection theory which opened the door to a resolution of these perplexing issues and to a modern understanding of antibody production (Ada, 1989). The crux of Burnet's theory was that in the animal there exist clones of cells which carry on their surfaces different recognition molecules. These molecules behave in a lock and key fashion, and bind the appropriate antigen. The reaction of the antigen with the recognition sites then activates that cell from a much larger population of clonal precursors, in effect selecting it. The selection process, through some unknown mechanism, then propels the cell down a long cascade of division and antibody production.

Burnet's proposal met with opposition from the start, and its vindication would require the advent of molecular approaches to immunology. As is so frequently true in the history of science, its final acceptance required concessions from both the germ line and somatic mutation factions, who lined up on either side of the debate (Silverstein, 1985). For the purposes of this discussion I will not follow the tortuous route through the next years that brought about our modern understanding of the mechanism of antibody diversity. During this intervening period of almost four decades, hundreds of thousands of years of investigators' time was spent working out the fine details of the immune system, which I can present here only in broadest outline. Although many details in immunology remain to be resolved, the mechanism of antibody diversity is well understood.

Antibody molecules share a common general structure, that of a "Y," with the portion that combines with the antigen being the ends of the prongs of the "Y" (Figure 1).

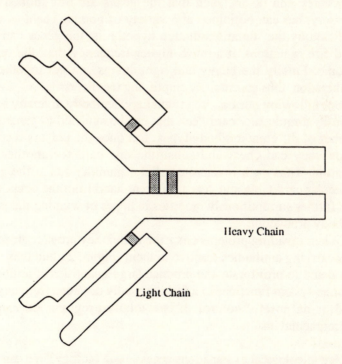

Figure 1. Four polypeptide chains, two light and two heavy, are joined together by disulphide bonds to form the antibody structure.

Each antibody molecule consists of two smaller units called light chains and two larger units known as heavy chains. The four submolecules are held together to form the complete, functional antibody molecule. Within the antibody-producing cell a number of genes are actually pulled out of their original position in the DNA and joined together to make the particular antibody molecule. There are four different families of genes, designated V, D, J, and C. A representative of each of these families will be extracted to form a composite gene in which a unique combination is assembled, like a hand at cards.

Antibody diversity arises at several levels. First, multiple variable genes are encoded in the germ lines, as had been predicted by the germ line theory, but there are only a few hundred at most to select from. Then a joining process takes place between the V, J, and D genes, which adds an additional level of variety. Next a recombinational inaccuracy can occur, such that the genes are not spliced together precisely, but can be joined at a variety of points. Then, as had been predicted by the somatic mutation hypothesis, the genes can undergo rapid fire mutations at a much higher frequency than the rest of the genome. Finally the heavy and light chains can get together in any combination, thus enormously amplifying the diversity.

So, following our Las Vegas analogy, antibody diversity is brought about by a molecular card deck that can be shuffled to generate a vast number of different antibody types. But once the players receive their hands, they can cheat and substitute one card for another through mutation. Thus each inveterate cellular gambler sits at the green felt molecular card table and gets a different hand from the deck, which he then further surreptitiously modifies in hopes of winning the jackpot of antibody production.

When invading proteins enter the body, they are recognized by the cells carrying antibodies reactive to them (Figure 2), and these cells are stimulated to proliferate and produce large quantities of antibody. The immune system functions in a fashion totally divorced from any guiding hand or purposeful control. It is a totally programmed response to environmental insult.

Are teleological explanations necessary in biology? We can see from this brief recounting of the development (or should I say evolution?) of

Antigen		Antibody

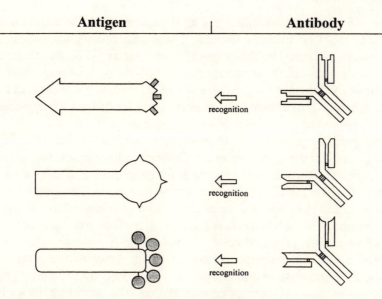

Figure 2. Antigens have characteristic regions, known as epitopes--molecular shapes that vary from antigen to antigen. Through recognition of antigen shapes, the generation of the appropriate antibodies is selected.

a materialistic, mechanistic explanation for one of biology's cornerstones two important lessons for our consideration of the Darwinian debate; first, that selection determines the immune response, and second, that it is unnecessary to invoke teleological explanations to account for the phenomenon of antibody diversity.

In developing theories of the immune system, no less a biologist than the great Paul Ehrlich, founder of modern immunology, used the phrase *"uralte protoplasma Weisheit,"* the "ancient wisdom of the protoplasm" (Silverstein, 1985). Ehrlich was proposing his side chain theory of antibody formation, and this ill-chosen term reflects, I believe, ambiguity that early biologists felt in designing mechanistic theories with which they were uncomfortable. Those early models were much too simple to account for the incredible complexity of biological systems. Ehrlich wrote the line in 1897, however, and in view of the vast gaps in his understanding, it is hardly surprising that teleologic phraseology crept into his descriptions.

At the time Ehrlich proposed his theory, it was thought that the only immunological response was to pathogenic organisms and toxic substances. These were believed to be few in number and thus the total number of different antibodies that an organism would be required to generate would be severely limited. It was unnecessary to propose a vast amount of unused antibody specificities. But as the immunological repertoire expanded to unmanageable proportions, it seemed that a Darwinian explanation for the existence of the immune system was no longer tenable. This ushered in the germ line theory with its immense and largely unused cornucopia of antibody types.

The guiding principle of Darwinian evolution convinced many investigators that the germ line theory, at least in its simplest form, had to be incorrect. It would make no sense for an organism to carry a vast amount of information that would never be employed. Surely it would be eliminated by natural selection. The development of our understanding of antibody diversity is a perfect example of the predictive power of the theory of evolution. If investigators had accepted a guiding, purposeful hand in the molding of the immune system, then the vast immune repertoire proposed by the germ line theory makes sense. It requires no explanation. The germ line theory is perfectly satisfactory, since the creator could look into the future and know what antibodies would be required.

But attention to the principle of natural selection forced a rejection of the germ line theory and a search for new experimental data. The present-day synthesis incorporating our knowledge of gene splicing, mutation, and rearrangement of the antibody-forming genes provides us with a completely rational and suitable account.

Thus in the development of immunological theories, the idea of purposeful design is no explanation at all, and is simply an excuse for muddled thinking.

If there is no scientific reason to evoke purpose in biology, does it have any place at all in science? Freeman Dyson (1979), a professor of physics at Princeton, has discussed the philosophical implications of his work in *Disturbing the Universe*. He has developed an interesting argument for a sort of purposeful spirit in the workings of the universe at the subatomic level. I think that this is a comforting notion, and is wholly consistent with the facts. Dyson, however, along with the vast majority of scientists, sees no reason for introducing the idea of purpose into biology.

The last hundred years and the work of thousands of investigators have established over and over again that mechanistic explanations are entirely adequate to explain the existence of the living world.

Descriptions based on mysticism, divine intervention, or purposeful guidance are untestable and provide no basis for an understanding of biological mechanisms. Purpose is a hindrance when introduced into discussions of biological phenomena, and can actually confuse us and impede our search for the truth.

NOTE

[1]Dobzhansky's statement is one of the most oft-quoted assertions in biology, and is cited in Futuyma's book, among many other sources.

REFERENCES

Ada, G. L. 1989. The concept and birth of Burnet's clonal selection theory. IN: *Immunology*: 1930-1980 (Ed. Mazunder). Wall and Thompson, Toronto. pp. 33-41.

Dyson, F. 1979. *Disturbing the Universe*. Harper and Row, New York.

Futuyma, D. 1982. *Science on Trial*. Pantheon Books, New York.

Kindt, T. J., and Capra, J. D. 1984. *The Antibody Enigma*. Plenum Press, New York.

Lederberg, J. 1959. Genes and antibodies. *Science* 129:1649.

Mazunder, P. H. 1989. The template theory of antibody formation and the chemical synthesis of the twenties. In *Immunology*: *1930-1980* (Ed. Mazunder). Wall and Thompson, Toronto. pp. 13-33.

Silverstein, A. M. 1985. History of immunology: A history of the theories of antibody formation. *Cellular Immunology* 91:263-283.

10a
Response to K. John Morrow, Jr.
Michael J. Behe

THE CONCLUSION THAT Professor Morrow draws from the history of theories of antibody diversity is a splendid, shimmering, jewel-like example of the ability of a theory, here materialistic evolution, to supply "facts" to the true believer that the mere, neglected, primary data in no way warrant. The believer looks upon the most innocuous facts and sees in them a stunning confirmation of his theory, where a person who is not committed to the hypothesis sees irrelevant or, sometimes, hostile information. Thus the believer builds a great edifice of pseudo-knowledge which, like cotton candy, is spun from a little bit of sugar and a lot of air. Let us, then, try to deconstruct Professor Morrow's argument, separate the solid material from the airy hypothesis, and see if anything is left to build on.

The ability of the immune system to respond to an invasion by virtually any foreign substance is truly amazing and, as Professor Morrow has told us, until recently it had puzzled scientists. An understanding of what such a system might look like was hampered for a long time by a lack of data, which is almost always the case in science. So, in the absence of sufficient relevant data, scientists tried to come up with an explanation for antibody diversity in terms of concepts that were available at the time. Thus, since it was known that antibodies were proteins and that proteins were flexible polymers, it was speculated that, well, maybe antibodies wrap themselves around a foreign substance and somehow freeze into that shape, somehow. Or maybe, since proteins were known to be coded by genes and since the mammalian cell had enough DNA to code for a lot of proteins and since host defense is so vitally important, well, maybe there is a very large number of such genes. Or maybe, since it was known that changes, mutations, in genes could occur, maybe somehow the mutation rate for cells that produce antibodies is cranked up very high, so that many different types of antibodies could be produced without requiring coding in the DNA. Or maybe it was a combination of some of these. Or maybe it involved something no one suspected.

As it turns out, the explanation of the basis of antibody diversity had to await a startling discovery: genes coding for proteins often do not occur contiguously—essentially they are genes in pieces. This means that, for example, the piece of DNA that tells how to make the lefthand portion of a given protein can be separated from the piece that tells how to make the middle portion, which can be separated from the piece that tells how to make the righthand portion. In the cellular process that "reads" the information, the disparate messages are joined together and a single, continuous protein chain is produced.

It was subsequently seen, as Professor Morrow has mentioned, that genes coding for antibodies are generally broken into four pieces: the V, J, D, and C regions. Now, there are a number of copies of each of these regions, differing one from the other, in the germ line of an individual. When cells that make antibodies develop in the body, a clever trick is employed. Instead of one set of V, J, D, and C pieces always joining together, as is the case for the pieces of most other proteins, any V can join with any J and they can join with any D and any C. This increases the number of combinations fantastically. Let us assume, for the purpose of argument, that there are 100 different V regions, 100 different J regions, etc. Then the number of different combinations of V and J are 100 times 100, and the combinations of V, J, and D are 100 times 100 times 100, and the combinations of V, J, D, and C are 100 to the fourth power, which is one hundred million. So a very large number of combinations can arise from a very limited number of genes. This is the currently accepted explanation for the generation of antibody diversity.

Professor Morrow has indeed selected a very elegant biological system to discuss. He has shown us that the mechanism for the production of antibody diversity is very clever, that complexity is generated from a finite number of components, that this mechanism is for all practical purposes able to deal with virtually any foreign material in the body, and that to a large degree it is through the efforts of science that we have come to understand how this system works. What Professor Morrow has *not* told us, however, is how he knows that such an elegant system evolved in a nondirected manner.

The question of what route the evolution of such an intricate system might have taken is also left unaddressed. Instead his argument sets up several straw men which he then proceeds to knock down. The first straw man set up to serve Professor Morrow is the notion that a materialistic, mechanistic explanation for the *functioning* of a biological system, as opposed to such an explanation for its *origin*, is a refutation of the idea of purpose. To my knowledge no one at this conference has advanced the idea that, say, the foot, the heart, or the digestive tract operate in anything other than a mechanistic fashion. But plenty of people profess to see purpose in their functioning. The fact that the immune system is a mechanism says absolutely nothing about whether it has a purpose.

The second straw man found in the preceding talk is the notion that one of the original hypotheses on antibody diversity, the germ line hypothesis, was somehow put forward to save teleology, as if the hypothesis was popular mostly in seminaries and Bible colleges. In fact, even a cursory look through biochemistry textbooks of recent years and the original literature shows a distinct lack of religious allusions when the germ line theory is discussed. It is difficult to guess why that hypothesis should bear the burden of teleology.

In the final analysis, then, the only positive argument that Professor Morrow advances to defend the idea that the system for generating antibody diversity actually evolved is theological: God wouldn't have done it that way. Professor Morrow informs us that the germ line model was the one God would have selected, but that the germ line model is incorrect. Therefore, one infers, God does not exist or at least does not mess with biology. Materialistic forces are all that are left to get the job done, so nondirected evolution must have produced antibody diversity.

But why is the germ line model approved by the deity? Well, because, "the creator could look into the future and know what antibodies would be required." The theological insight in this statement is breathtaking! By the same reasoning we can know that our hands are strong evidence for evolution since the creator could look into the future and see that we would eventually need bottle openers, scissors, and corkscrews, and would therefore have appended Swiss army knives to the ends of our arms. With his keen appreciation for theological issues Professor Morrow apparently has decided that a creator would not provide flexible, multipurpose, *clever* tools to his creatures.

Regrettably, I have no formal theological training, so Professor Morrow's argument is wasted on me. The only advanced work I have done is in science and I must rely on results from that discipline to reach a conclusion. If I may, then, let me return to the immune system and suggest some questions that science can at least in principle address, though perhaps not at the present time.

One definition of "evolution," according to Webster's Ninth Collegiate Dictionary, is "a process of continuous change from a lower, simpler, . . . to a higher, more complex state." Certainly Professor Morrow would agree that Darwinism must hold that the immune system of the higher eukaryotes did not come suddenly into being, but that it must have developed gradually from a simpler system. What difficulties would that development have to overcome? The first problem, of course, is the origin of life, but we will pass over that here.

The second difficulty is the origin of splicing. How at first did different regions of a nascent "message" get hooked together in the proper order and the intervening sequences edited out, and, more important, how likely is that mechanism, which in the modern world requires many different proteins, to have developed through a nondirected search?

The next hurdle is inserting the antibody into the cell membrane. But even then a cell with an antibody in its membrane is useless unless there is a specific feedback system simultaneously to cause the cell to proliferate and to begin exporting soluble antibody. This will require another two to three proteins. In this respect a cell with just an antibody would be like a person holding a steering wheel, looking for a car to hook it on. If all of these difficulties are eventually overcome we have a population of cells exporting antibodies that bind to a given antigen. But, as stated in the popular textbook *Biochemistry* by Voet and Voet (page 1114), "Antibodies, for all their complications, only serve to identify foreign antigens. Other biological systems must then inactivate and dispose of the intruders." To an invading agent, being bound by an antibody is like being shot with a dart gun: the dart may stick, but it does no harm. Thus after an organism has gone to all the trouble of developing a diverse array of antibodies that can recognize many foreign bodies, it is still virtually helpless. In modern-day organisms, after intruders are identified by antibodies, a completely different pathway called the "complement system" actually kills or gets rid of the invader. The complement system is a highly regulated pathway of more than a dozen different proteins. How did the interdependent antibody and complement systems develop, step-by-step; what was the selective advantage to the organism for each step along the way; and what are the odds that such an intricate system would develop in a nondirected search?

As I stated earlier, questions such as these are within science's sphere of competence, and until the questions are answered in their favor *by experiment*, proponents of nondirected evolution have no right to cite the fantastic, intricate, clever immune system as evidence for their views.

A number of proponents of naturalistic evolution are fond of saying that "the theory of evolution is about as much in doubt as that the earth goes around the sun." They seem not to notice that the proposition that the earth does go around the sun, although once controversial on religious grounds, is now universally accepted. Nondirected evolution, however, remains as controversial as when it was first proposed. The reason for these different receptions is that strong evidence has been produced by science to support the solar system, but *convincing evidence of the truth of Darwinism has not yet been produced.* Until such evidence is produced, no theological, philosophical, or cotton candy arguments will quell the controversy. Until then every person has the right, on solid scientific grounds, to regard Darwinism as an interesting but very doubtful hypothesis.

10b
Reply to Michael J. Behe
K. John Morrow, Jr.

I AM PLEASED TO HAVE an opportunity to respond to Dr. Behe's criticism of my paper, as it gives me an opportunity to elaborate on some aspects of my talk, and clarify what I believe are misstatements or misinterpretations of my position. As Dr. Behe has accused me of raising straw men, I could perhaps point to a few of his own. This discussion could perhaps be subtitled, My straw man can beat up your straw man."

The thrust of Dr. Behe's criticism is that in arguing the validity of Darwinian evolutionary theory, a comparison of the various hypotheses of immune diversity is (1) irrelevant and (2) begs the issue of where the immune system came from, for which he thinks there is at present no adequate answer. Let me address these issues one at a time.

In the first place, I chose this topic because it is restricted enough and sufficiently well investigated to be understandable in the framework of a short presentation. I did not wish to deal with the evolution of the immune system, since it is an extremely complex topic that does not lend itself to a didactic exposition. But since Dr. Behe has brought this up I will be happy to do so.

But first, let us consider his criticism that my analysis of the various hypotheses of immune diversity constitutes a straw man, which I use to buttress my own narrow and unsubstantiated belief in evolutionary dogma. Perhaps we can best do this by turning the argument around.

Say the germ line theory *had* proven to be true, and say that a mammal contained a huge amount of genetic information coding for millions and millions of different antibody types. Most of this information would never be marshaled, either in the individual's lifetime or the lifetime of the entire species. Say, further, that it were discovered that mammals carried genes that coded for man-made compounds that had been synthesized only this year, and that bore absolutely no resemblance to any naturally occurring compound. Do you think that a creationist would argue that such a system constituted scientific evidence for purpose in biology? And do you think, moreover, he would argue that the existence of genes of no selective value to the individual (or perhaps to the entire species) would suggest that mechanisms other than Darwinian evolution and natural selection guide the development of complex biological system? I leave it to the reader to decide this question.

Dr. Behe points out quite correctly that the enigma of immune diversity was solved using molecular techniques that were not available when the question was formulated. It was not resolved through a deductive process carried out by monks speculating in the confines of their cells in a lonely mountainside monastery. I agree with the statement that immune diversity was never a topic of theological conjecture. My point was that in the early years of the twentieth century there were several competing theories to account for how antibodies were produced. The theory of natural selection predicted that the germ line hypothesis could not possibly have been right, at least in its simplest formulation. This is a characteristic of a good scientific theory; it makes predictions, generates experiments, and drives the field forward.

Dr. Behe accuses me of presenting a biological mechanism and then arbitrarily stating that God wouldn't use it, and therefore God does not exist. *Au contraire.* My point was that what we know about the immune system gives us no reason to propose any kind of divine purpose or conscious will in shaping it. Clearly God may exist and he may use any means he wishes to shape the universe. Perversely, he may have created the universe ten minutes ago, and everything in it may be placed there simply to confuse and distract us. I certainly don't believe in such a wicked god, and neither does Dr. Behe or anyone else who defends Christian theology. The divine purpose that we are considering here finds its roots in biblical accounts of God and his powers. This God is a knowing, loving deity.

But Dr. Behe's obstruction is typical of anyone who clings steadfastly to the notion that living systems were shaped by a knowledgeable, rational, purposeful consciousness. When one presents a biological system that doesn't fit these criteria, Dr. Behe can always say, "You are trying to put ideas in God's head. Neither you nor anyone else knows what God was thinking when he created living systems."

If there is a rational purpose to life and if Darwinian evolution is inadequate to explain it, then the divine hand that shaped it must have created us in his image and his thought processes must be similar to our own. This is what the God of Christianity is all about. He is not arbitrary. He is not wicked. He is not capricious.

My point is that an inspection of the history of biology shows that, over and over again, theories of life requiring a teleological basis have been found, in the light of modern molecular investigations, to be unnecessary. This realization certainly doesn't argue that God does not exist, but simply that it is unnecessary to propose a divine purpose to explain how living creatures came to their present state of development. But this is precisely where Dr. Behe and I differ. He says that the most important materialistic, mechanistic principle of biology, the theory of evolution, is inadequate to explain the diversity of living systems, and therefore we must evoke divine purpose, and a divine purpose that intervenes on a step-by-step basis.

But to return to Dr. Behe's second criticism of my argument, are there really unresolvable questions concerning how the immune system could have evolved solely through natural selection? Most of Behe's specific points go back to the notion that the evolution of complex systems is impossible because none of the individual components has selective value independent of a complete, integrated whole. But this is a straw-man argument, which implies that unless the system is working flawlessly it won't work at all. Clearly in Dr. Behe's universe there is no Model T on the way from the horse and buggy to the Mercedes.

For instance, Behe says that an antibody molecule by itself would have no selective value for an organism, and therefore it is impossible that the immune system arose through a gradual process of evolution and natural selection. He then goes on to mention that a specific protein, known as complement, must be present in order for immunoglobulin molecules to exert toxicity against foreign invaders; he suggests that without a fully developed complement system antibodies would be of no selective value. But this statement ignores the fact that various primitive complement-like molecules have been identified in invertebrates, which function in a nonspecific manner to protect the organism from foreign invaders. In fact, complement-like molecules have been found in a variety of protostome and deuterostome invertebrates. These molecules are used for triggering phagocytosis (a process by which pathogens are devoured), possibly through the intervention of other molecules known as lectins.

Behe seems oblivious to the evidence that antibodies are members of a class of molecules, the immunoglobulin super family, many of whose members are involved in cell-cell recognition and signaling. In fact, super family molecules of the class known as β-microglobulin exist on the brain of the squid. It has been suggested that the immunoglobulin superclass evolved from a common precursor responsible for mediation of cell signals. Members of the Ig super family are found in the membranes of neurons, indicating that molecules that communicate signals use a common basic plan, modified for individual tasks. Thus a common molecular precursor has been revamped over and over again in the course of evolution. The ancestral immunoglobulin molecule no doubt functioned as a crude recognition molecule, allowing some degree of reactivity against foreign pathogens. From this vantage point we can think of the immune system as being one component of a broad-ranging communication network, occurring throughout the phylogenetic tree and throughout different organ systems within individuals.

It is not known how an internal recognition system became modified in the course of evolution so as to "turn outward" and recognize foreign proteins. But it does not take a great leap of the imagination to conceive of a primitive species that would use immunoglobulin-like molecules for recognition of foreign proteins through a mechanism of broad specificity, eventually developing through natural selection a panel of recognition signals, and then later evolving a broader and broader collection of molecules through gene duplication. Eventually mutations destabilizing some of these recognition genes might appear. These unstable regions might generate a wider range of molecules with a weak affinity for pathogenic organisms.

One can imagine what an advantage an organism would have with this crude proto-recognition system for the identification of foreign marauders, even if it were initially extremely inefficient.

A reasonable idea does not constitute a proof. I have suggested these possibilities simply to illustrate how artificial Behe's argument is, and to emphasize that there is nothing arcane or mysterious about the evolution through intermediate steps to a highly sophisticated biological mechanism. I would have to admit that we really do not know precisely how the immune system came to be. But Dr. Behe's objection to a Darwinian evolutionary interpretation is precisely what troubles me about the anti-evolutionist stance; because we don't have an ironclad proof of every precise feature of the evolution of the immune system in hand, Behe suggests that we just abandon the whole enterprise and accept a fuzzy and ill-defined teleology as a scientific explanation.

This strikes me as a profoundly unscientific approach that is foreordained to failure.

11

X Does Not Entail Y:
The Rhetorical Uses of Conflating Levels of Logic
Arthur M. Shapiro

> Science, to put it bluntly, is uneasy with beginnings.
> Mythology, on the contrary, is concerned above all with
> what happened "in the beginning." . . . Its signature is
> "Once upon a time," . . . But it differs most importantly
> from science in that its explanatory account of how we
> began is also a prescriptive account of how subsequent
> beginnings . . . should proceed; the Last Supper, for
> example, tells us not only how the Christian era began
> but how its energies can be periodically renewed. . . .
> • Dudley Young, *Origins of the Sacred:
> The Ecstasies of Love and War* (1991).

> Facts are precisely what there is not, only
> interpretations.
> • F. Nietzsche, *The Will to Power.*

IT'S A GREAT PLEASURE to be here in Texas, where a certain regional
neo-Populist politician is fond of saying there's nothing in the middle of
the road but yellow lines and dead armadillos. I beg to disagree, at least
insofar as the middle of the road applies to the relationships between
religion and science.

Andrew Dickson White called his seminal work *A History of the
Warfare of Science with Theology in Christendom.* No one can deny
that such warfare has occurred, and even its historical necessity may not
be in dispute. Like literal wars, much of that warfare resulted from the
artful promotion of misperception and misunderstanding by the pro-war
parties, usually on both sides. To recognize such manipulation offers at
least a glimmer of hope that the hostilities may at last come to an end.

We are here to focus on Darwinism, or neo-Darwinism, because that
is the target Phillip Johnson has chosen to attack. I am going to dispute
the uniqueness and even the appropriateness of that target. I maintain
that the concentration on evolution as the point where science and
religion collide is both a product of historical forces and of the kind of

rhetorical manipulation that gets people and nations into wars. I further maintain that the dichotomies employed by the propagandists of the science-religion war are false, and that proper analysis of the logical structure of their argument rapidly destroys its validity.

At the same time, I want to stress two caveats. First, this is intended as a polemic, not as a formal axiomatization of the argument. Second, I do not claim that anything new is to be found here. My arguments are old; perfectly adequate specimens of them, fashioned in fact to deal with the question of Darwinism, can be found in sources from the nineteenth century, some of which I will cite.

Here is my argument in a nutshell: Biological evolution (Darwinism, neo-Darwinism) entails no particular position on the ultimate origins of either life or the universe. Evolution is a subject studied by the methods of science. To conduct scientific investigation per se entails no claim to intellectual hegemony or ontological priority over other potential "ways of knowing." The contrary claims, implicit or explicit in the arguments of both theists and atheists, flow from a conflation of evolution with evolutionism or of science with scientism (or positivism, or materialism, or some other ism). The conflation may be pertinent to discussions of human affairs and society but at the same time is obfuscatory and logically invalid, as conflation by definition is.

I further argue that biological evolution is no more inconsistent with religion than are other sciences, and that the attacks on it specifically are best understood in sociological and political, not philosophical, terms.

* * *

Phillip Johnson is a very intelligent man, highly skilled in argumentation. It annoys me to find him making the same vulgar errors as appear routinely in the presentations of professional creationists. I refer, of course, to the conflation, if not equation, of science and scientific disciplines with ideological positions. But this is no new error. Several years ago I had the opportunity to teach a series of Sunday school classes on creation and evolution to a group of adult Presbyterians, and I was able to find outstanding specimens of debate in the nineteenth century Presbyterian literature to present to them. Charles Hodge (1797-1878) was sixty-two years old when the *Origin of Species* appeared and was firmly established as a rock-ribbed conservative theologian; he boasted that no new idea ever originated in the Princeton Theological Seminary, which he dominated for decades. His book *What Is Darwinism?* (1874) declares:

This is a question which needs an answer. Great confusion and diversity of opinion prevail as to the real views of the man whose writings have agitated the whole world, scientific and religious. If a man says he is a Darwinian, many understand him to avow himself virtually an atheist; while another understands him as saying that he adopts some harmless form of the doctrine of evolution. This is a great evil.

It is obviously useless to discuss any theory until we are agreed as to what that theory is. The question, therefore, what is Darwinism? must take precedence of all discussion of its merits.

I commend Hodge's formulation and analysis to my readers. Referring to Darwin, he says:

... he uses the word natural as antithetical to supernatural. Natural selection is a selection made by natural laws, working without intention and design. ... In using the expression Natural Selection, Mr. Darwin intends to exclude design, or final causes.

and:

There are in the animal and vegetable worlds innumerable instances of at least apparent contrivance, which have excited the admiration of men in all ages. There are three ways of accounting for them. The first is the Scriptural doctrine, namely that God is a Spirit, a personal, self-conscious, intelligent agent. ... This doctrine does not ignore the efficiency of second causes; it simply asserts that God overrules and controls them. ...

The second method of accounting for contrivances in nature admits that they were foreseen and purposed by God, and that He endowed matter with forces which He foresaw and intended should produce such results. But here His agency stops. He never interferes to guide the operation of physical causes. ... This banishing God from the world is simply intolerable and, blessed be His name, impossible. An absent God who does nothing is, to us, no God.

The third method of accounting for the contrivances ... refers them to the blind operation of natural causes. This is the doctrine of the Materialists, and to this doctrine, we are sorry to say, Mr. Darwin, although himself a theist, has given his adhesion. ...

And to summarize:

The conclusion of the whole matter is, that the denial of design in Nature is virtually the denial of God. ...

We have thus arrived at the answer to our question, What is

Darwinism? It is Atheism.

No, it is not.

Observe how easily Hodge implicitly equates Darwinism with Materialism. On reflection, the logical leap he makes, from Darwinism to Materialism to Atheism, becomes increasingly problematic. Darwin's sin, it appears, was to fail to acquiesce to the claim that "God did it." Some theologians indeed might argue that to seek alternative, purely materialistic explanations for phenomena already attributed to God is to commit the sin of pride. Darwin's great intellectual triumph was to provide a mechanism, natural selection, that could account for what Hodge calls "contrivances": to provide the appearance of design without invoking a Designer.

Clearly, the availability of such a mechanism would gladden the heart of anyone who for whatever reason wished to banish God from the universe. But it itself could not banish God from the universe. If the mechanism works, if it is proven valid and sufficient, then it renders God simply redundant in that context. But that is not to say that it disproves his existence; it merely makes it a teensy bit less necessary for explanatory purposes. Hodge dismisses deism with a wave of the hand: "An absent God who does nothing is, to us, no God." A deist might well reply that a God who merits belief only insofar as he is necessary as a unique explanation of biological phenomena is a pretty weak God.

Somewhere years ago I saw a cartoon showing an upset wife in her scientist-husband's lab. The white-coated husband is fiddling with test tubes, and she says, "But if you reduce us to molecular-level phenomena, what happens to our mystique?" It is no secret that the phenomenal growth of science in the last few centuries has been largely at the expense of both religion and secular philosophy: both have had less and less to claim as their unique explanatory domains.

There are various reasons why materialistic explanations in biology have often been viewed as more threatening to religion than those in physics and chemistry. But biological explanation, including Darwinism and neo-Darwinism, has no unique intellectual flaw that renders it particularly weak as an alternative to religion. This is, however, the impression conveyed by many of its critics, ranging from the folks at the Institute for Creation Research to, say, Phillip Johnson. More on this anon.

* * *

Critics of evolution, including Johnson, persistently conflate

Darwinism or neo-Darwinism with the origin of life or of the universe, or both. There is simply no entailment here. Darwin was in no position to devise tenable hypotheses about the origin of life, since the basic biochemistry of life was unknown. There is nothing wrong with such hypothesizing, of course. It is perfectly normal scientific procedure to attempt to generate materialistic explanations for what appear to be material phenomena. We need to be clear as to the logic of entailment here. The validity or invalidity of Darwin's, or anybody else's, materialistic-mechanistic hypotheses about the evolution and diversification of life implies nothing about the validity or invalidity of any given suggestion about the origin of life, let alone the universe—except insofar as true explanations at those two levels could not be mutually contradictory as to, for example, the basic chemistry of life. That is, origins-of-life hypotheses must account successfully for the origins of the biochemicals that define life as we know it, or of plausible precursors of those, not some utterly different set unrelated to life on earth today.

But a successful creation-of-life experiment in the laboratory would merely demonstrate that such a thing was possible. It would not, and could not, prove that that was how it happened the first time. The first synthetic test-tube birth cannot logically dictate the death of God.

The Epilogue of Thaxton, Bradley, and Olsen (1984) reviews and criticizes theories of origins, concluding with a plea for "metaphysical tolerance." These authors correctly, I think, compartmentalize origins as separate questions from biotic evolution. They draw a distinction between "operations science," the "second causes" of Hodge, and "origins science." Their point is that hypotheses about singular events, such as the origin of the universe or the origin of life, cannot be falsified; this, they believe opens the door to inclusion of the supernatural as a "scientific" explanation. I shall return to this theme too, noting in passing the intellectual bankruptcy of panspermia as an alternative explanation for the origin of life. Panspermia merely puts the question of origins a step back into deeper unknowability somewhere in space, with both materialistic and theistic scenarios remaining viable but somewhat further away.

* * *

Science is not an ideology. Scientism and perhaps positivism are. As a philosophical position, positivism is in advanced decay, and for anti-Darwinists to continue to flog it is either a rhetorical ploy or a demonstration of being out of touch. But that is not to say that the doing of science is independent of certain minimal and intrinsically ideological suppositions. A large literature can be found on this point but because I have no time here to review it, I will give my favorite presentation of those suppositions—which happens to be my own (Shapiro, 1987):

> One is able to do science at all only if one accepts certain intrinsically unprovable postulates about the universe: that a material universe exists in some meaningful sense; that the evidence of reason and our (extended) senses is sufficient to comprehend that universe; that the universe is lawful; and that its laws are and always have been the same everywhere. . . . This is a materialistic belief system, an ideology if you will, no more subject to empirical or logical validation ("proof") than any religious belief system.
>
> Note that the minimum set of materialistic beliefs enumerated above neither denies nor excludes the possibility of the supernatural; it *ignores* it. . . . Science *per se* neither affirms nor denies the existence of God. This was perfectly clear to the liberal Presbyterian theologian James Woodrow when he made his famous defense of Darwinism in 1884, and it is no less clear now.

In my Presbyterian Sunday school class I rebutted Hodge with a reading from Woodrow. Woodrow to a degree conflates origins with evolution too, but he (I think correctly) keeps his logic of entailment straight nonetheless:

> [The] definition now given [of evolution], which seems to me the only one which can be given within the limits of natural science, necessarily excludes the possibility of the questions whether the doctrine is theistic or atheistic, religious or irreligious, moral or immoral. It would be as plainly absurd . . . to inquire whether [it] is white or black, square or round, light or heavy. In this respect it is like every other hypothesis or theory in science. These are qualities which do not belong to such subjects. The only question that can rationally be put is, Is the doctrine true or false? If this statement is correct—and it is almost if not quite self-evident—it should at once end all disputes not only between Evolution and religion, but between natural science and religion universally.

And:

> To prove that the universe, the earth, and the organic beings upon the earth, had once been in a different condition from the present, and had gradually reached the state which we now see,

> could not disprove or tend to disprove the existence of God or
> the possession by Him of a single attribute ever thought to
> belong to Him. . . . He is as really and truly your Creator,
> though you are the descendant of hundreds of ancestors, as He
> was of the first particle of matter which He called into being, or
> the first plant or animal, or the first angel in Heaven.

Scientism or atheism has agendas that leave no room for God. It can
be claimed that the distinction between science and scientism, or
evolution and evolutionism, may be valid but nonetheless unimportant,
insofar as the effect of science or evolution is to promote the agendas of
scientism or evolutionism. That is, any successful scientific idea by
definition is another nail hammered into the coffin of theism, insofar as
public perception is concerned. The public is no more sophisticated in
these matters than are the theologians, philosophers, biochemists, and
law professors who confuse these terms and concepts. And the social
consequences of science-driven, mass apostasy are too horrible to
contemplate. (Rachels, 1991, ruminates on potential bases for post-
Darwinian morality.)

I do not like functionalist arguments, on principle. I remember too
vividly that Vavilov's pursuit of theoretical questions in agrogenetics
was sufficient to define him as a German agent in the minds of the
Stalinist authorities. The USSR, it will be recalled, had an entire
criminal category, "wreckers," those who sabotaged the cause of the
People by failing to be duly subservient, and thus served the cause of
the enemy.

Of course it is naive to claim that ideas do not have consequences.
They do. By giving God less and less to do, science has not been good
overall for theism. The single-minded focus on Darwinism and neo-
Darwinism, however, in itself betrays the functionalist reasoning and
motivation of so many of the critics. This symposium is not the place to
do it, but it would be an interesting exercise to attempt a formal
comparison of the intellectual structure of, say, neo-Darwinism and the
theory of plate tectonics. (I admit that attempts to date to axiomatize
evolutionary theory have been rather unsatisfying.) It seems to me that
the structure of conjecture and extrapolation is very similar and perhaps
effectively isomorphous in these two fields (which, by the way, are not
only complementary but are strongly mutually reinforcing) and that if
the alleged intellectual weaknesses of evolutionary theory were really
being pursued for their own sake, those of structural geology would
have merited comparable attention. Despite the existence of a bumper

sticker that says "Stop continental drift" there seems little concern for the moral implications of subduction. And if the critics are right, *there should be*, because banishing the deity from shaping the contours of the earth should be at least as damaging to his reputation as denying his responsibility for the shapes of moths' wings.

In practical terms, however, opposing science across the board does not work; science provides too many tangible comforts. The perception of practical reward from Darwinism is low ("operations science" can continue to support medicine without postulating phylogeny) and the emotional resonances attached to perceptions of our own uniqueness are high. For anti-Darwinians, then, conflation is a useful rhetorical tool.

But it is equally useful for ideologues of the other side, who claim that the successes of Darwinism indeed are nails in the coffin of God. One of the most articulate spokesmen for scientistic atheism in this country is Provine (1988). I pointed out to him that his arguments were very close to those of the conservative Presbyterian theologian, Hodge. He was not surprised; he said he came from a "long line of Presbyterian ministers" himself.

* * *

To admit that neither Darwinism, nor some eventual creation of life in the laboratory, nor the theory of plate tectonics, nor any other conceptual or empirical achievement of science can definitively settle the question of God's existence is an unappealing prospect to both theists and atheists. It bothers atheists because it leaves the door open to people (irrationally, as the atheists see it) continuing to cling to religion out of wishful thinking. It bothers theists because it obliges them to seek justifications for their belief which do not depend on the necessity of God to explain the material world. And that is hard work.

Both the humanities and the special, sort-of-hybrid discipline of philosophy of science have been wrestling with the nature of claims to objectivity. Phillip Johnson and others of our critics have accused us, with some justice, of acting as if we stand on high ground, free from ideological presuppositions. I have tried to show in this paper that we have certain basic ones, but that they are only a subset of those constituting a full-blown materialistic ideology.

With that said, a few closing words on the role of ideology and the realism-relativism question in science in general, with special reference to evolution, seem called for.

Alasdair MacIntyre (1988) says:

> There is no other way to engage in the formulation, elaboration, rational justification, and criticism of accounts of practical rationality and justice except from within some one particular tradition in conversation, cooperation, and conflict with those who inhabit the same tradition . . . . Considerations urged from within one tradition may be ignored by those conducting enquiry or debate within another only at the cost, *by their own standards*, of excluding relevant good reasons for believing or disbelieving this or that or for acting in one way rather than another. Yet *in other areas* what is asserted or enquired into within the former tradition may have no counterpart whatsoever in the latter . . . . there is no set of independent standards of rational justification by appeal to which the issues between contending traditions can be decided.

MacIntyre's view here is an echo of Kuhn's notion of incommensurable paradigms in science. By this view, theistic and atheistic perspectives are so utterly different as to share no common ground on which to communicate. A related notion is elaborated by the philosopher Rorty (1979), who seems to say that science cannot set its own rules for what is or is not scientific, because it is self-interested and therefore cannot stand on neutral ground.

I am not going to resolve such claims here, though I would hazard a prediction, hardly original, that such trendy stuff will persist for even a shorter time than did all of the once-fashionable attacks on scientific realism. (Only a few weeks ago Marjorie Grene declared on my own campus that "everyone is some kind of a realist now. What's the point of denying that a real world exists?")

It seems to me that the first step to understanding the intellectual structure of the origins problem is to disentangle carefully the levels at which entailments are claimed. Once that is done, it should be clear that it is unreasonable to require science to do theists' homework for them. Here is Phillip Johnson (1990) speaking:

> The absence of proof "when measured on an absolute scale" is unimportant to a thoroughgoing naturalist, who feels that science is doing well enough if it has a plausible explanation that maintains the naturalistic worldview. The same absence of proof is highly significant to any person who thinks it possible that there are more things in heaven and earth than are dreamt of in naturalistic philosophy.
>
> Victory in the creation-evolution dispute therefore belongs to the party with the cultural authority to establish the ground rules that govern the discourse. If creation is admitted as a serious

possibility, Darwinism cannot win, and if it is excluded *a priori* Darwinism cannot lose. . . . Creation-science is not science, said the [National] Academy [of Sciences], because "it fails to display the most basic characteristic of science: reliance upon naturalistic explanations. Instead, proponents of creation-science hold that the creation of the universe, the earth, living things, and man was accomplished through supernatural means inaccessible to human understanding."

Phillip Johnson, conservative theist that he is, here betrays himself as deeply confused about commensurability. He accuses science of being unfair in not admitting into its own camp a philosophical perspective radically different from its own, and then using that refusal to cement its social hegemony. He very nearly goes so far as to claim, with MacIntyre or Rorty, that science *has no right* to be taken seriously when it attempts to define itself, because it is so deeply self-interested.

I do not hear science demanding that religions employ experiment to demonstrate the existence of God. I have heard positivists say that the notion of God is nonsense because it can't be tested by experiment. That, thank heavens, is a passé argument, but in any case it is *at a different level of discourse*. And that's the point.

Huber (1991) argues that the erosion of the 1923 Frye rule opened the way to "junk science" and chaos in American liability law. The Frye rule held that expert witnesses were permitted only when "their testimony was founded on theories, methods and procedures 'generally accepted' as valid among other scientists in the same field." This is indeed a prescription for preserving orthodoxy within science, yet—as philosophers, historians, and sociologists of science never tire of telling us—science, perhaps alone among human institutions, has built into it mechanisms for self-correction and the promotion of change.

I do not think that Phillip Johnson's antipathy to Darwinism extends far enough that he would embrace a relativist, perspectivist, or nihilist notion of "science" that would open the door not only to the Creator but to astrology, orgone energy, pyramid power, and all of that (and if he did, conservative Christians, thinking functionalistically, would see him as a stalking horse for Satan).

Thaxton et al. (1984) miss the boat when they call for inclusion of God as a viable hypothesis in "origins science." What they should say is that "knowledge claims" can be divided into a variety of categories, and that claims about origins are beyond the realm of scientific "proof." In MacIntyre's terms, the contending traditions in the realm of origins require no "independent standards of rational justification," insofar as they have been talking at different levels. The problem between them is a pseudo-problem. And it is not a Darwinian problem.

Unless one ideologically insists upon biblical inerrancy—at which point I will go have a beer.

Acknowledgments

This paper was cross-fertilized by a fortuitous encounter with the thinking of Prof. Richard C. Sinopoli of our Department of Political Science, with whom I both agree and disagree but whose sense of quotable quotes I greatly admire. I also have benefited from regular infusions of wisdom from Norman DePuy.

LITERATURE CITED

Hodge, C. 1874. *What is Darwinism?* Scribner, Armstrong & Company.

Huber, P. W. 1991. *Galileo's Revenge: Junk Science in the Courtroom.* Basic Books.

Johnson, P. 1990. *Evolution as Dogma: The Establishment of Naturalism.* Haughton.

MacIntyre, A. 1988. *Whose Justice? Which Rationality?* University of Notre Dame Press.

Nietzsche, F. 1967. *The Will to Power.* Tr. W. Kaufman and R. J. Hollingdale. Random House.

Provine, W. 1988. Evolution and the foundation of ethics. *MBL Science* 3(1): 25-29.

Rachels, J. 1991. *Created From Animals: The Moral Implications of Darwinism.* Oxford University Press.

Rorty, R. 1979. *Philosophy and the Mirror of Nature.* Princeton University Press.

Shapiro, A. M. 1987. God and science. *Pennsylvania Gazette* 86(1): 47-51.

Thaxton, C. B., W. L. Bradley and R. L. Olsen. 1984. *The Mystery of Life's Origin: Reassessing Current Theories.* Philosophical Library.

Woodrow, J. 1884. Evolution. Reprinted in J. L. Blau, ed., 1946, *American Philosophical Addresses 1700-1900*, pp. 488-513. Columbia University Press.

Young, D. 1991. *Origins of the Sacred: The Ecstasies of Love and War.* St. Martin's Press.

11a
Response to Arthur M. Shapiro
X Does Implicate Y: Implication and Entailment in the Creation-Evolution Debate
William A. Dembski

ARTHUR SHAPIRO HAS JUST argued that X does not entail Y, where X is biological evolution of the Darwinian or neo-Darwinian stripe, and Y is any particular position on the ultimate origins of life or the universe. To this I offer a hearty Amen. But I also ask, So what?

As a mathematician I've had plenty of experience in the logic of entailment—every theorem is entailed by some relevant set of mathematical axioms. As a philosopher who works in the logic of conditionals, I'm aware how entailment works outside mathematical contexts. Entailment is the strongest logical connection by far. To say that X entails Y is to say that it's impossible for X to be true and Y false. Alternatively, Y is necessary given X. Thus, to say that X does not entail Y is to say that it is possible for X to be true and Y false. But since Shapiro leaves Y completely open-ended on the questions of origins, to say that biological evolution does not entail any account of origins, be it theistic, materialistic, or whatever, is simply to say that biological evolution is logically compatible with any number of positions on the origin of life and the universe.

Given what is meant by logical entailment—and this is the sense in which Shapiro is using it—I must agree with his claim. Moreover, Shapiro's claim has empirical support: individuals with widely divergent views on origins have made their peace with, or (if you will) have surrendered to, neo-Darwinism. Certainly this is true of "scientistic atheists" like Will Provine. But it is also true of notable theists like Richard Swinburne, who even when writing on teleology and design admits the central claims of neo-Darwinism:

> Complex animals and plants can be produced through generation by less complex animals and plants . . . and simple animals and plants can be produced by natural processes from inorganic matter.[1]

Suffice it to say, there is no *logical* impossibility reconciling neo-

Darwinism with a host of philosophical positions on origins.

So what? Suppose I place Al and Bob in a room, lock it, reopen it an hour later only to find Al lying on the floor in a pool of blood, with Bob standing over him holding a smoking gun. Denote this scenario by X. Let Y denote the claim that Bob shot Al. Does X entail Y? Well, no. Al may have been suicidal and shot himself. Bob tried to prevent this and is now holding the gun which he was too late in taking away from Al. Suppose, however, we know that Bob and Al are mortal enemies, and that Al has no suicidal tendencies. With this background knowledge, does X entail Y? Again the answer is *No*. There might be a trap door in the room. Perhaps an enemy of both Al and Bob used the trap door to enter the room, shoot Al, and then place the gun in Bob's hand so as to frame Bob.

If this story appears fanciful, if I appear to be veering from the path of common sense, it is because the logic of entailment cannot distinguish between the banal, the bizarre, and the ridiculous. It can distinguish only between the possible and the impossible. The circumstantial evidence for Bob's killing Al may be excellent. If a video camera in the room happens to record Bob shooting Al, there will even be direct evidence for Bob shooting Al. But no amount of empirical evidence will *entail* Bob shooting Al. Bob's double might actually have shot Al. Bob's enemy might have rigged the video camera so that it only appears that Bob shot Al. I am not suggesting that our reason for believing that Bob shot Al becomes inferior because no evidence can entail this claim. Entailment is simply too strong a logical notion to apply in most matters of fact. In particular, it is the wrong philosophical tool for investigating the relation between Darwinism and origins.

It is here that Shapiro and I part company. Shapiro argues, and I quote,

> Biological evolution is no more inconsistent with religion than are other sciences, and . . . the attacks on it specifically are best understood in sociological and political, not philosophical, terms.

Philosophy has a lot more to say about the relation between biological evolution and world views than Shapiro is willing to admit. To move from entailment to sociology is simply too abrupt a leap. It is more than a sociological fact that, and I quote Shapiro, "the phenomenal growth of science in the last few centuries has been largely at the expense of . . . religion." By concentrating on entailment and jumping from there to sociology, Shapiro has ignored the epistemological question of what implications exist between Darwinism and religion.

In philosophy, implication is a more general notion than entailment. The scenario of Al and Bob locked in a room together with some appropriate background assumptions would implicate that Bob had murdered Al, but it wouldn't entail that Bob had murdered Al. Implication includes entailment, and therefore addresses questions of possibility and necessity. But implication also addresses questions of uncertainty, partial evidence, and probability. X can implicate Y without X having to force Y to be true under all possible circumstances. X can implicate Y, X can be true, but Y might still fail. Any lawyer will appreciate this point. As an aside, let me mention that this is one reason why I appreciate Phillip Johnson's work of weighing neo-Darwinism in the legal balances. A strict logico-deductive argument will never settle the creation-evolution debate.

I've said that Shapiro ignores the epistemological question of what implications exist between neo-Darwinism and theology. This is true in that he admits no implication other than entailment. Nevertheless, without assigning it any epistemological weight, he does mention a significant implication. Commenting on natural selection Shapiro notes,

> Darwin's great intellectual triumph was to provide a mechanism, natural selection, that could account for . . . the appearance of design without invoking a Designer. Clearly, the availability of such a mechanism would gladden the heart of anyone who for whatever reason wished to banish God from the universe. But it itself could not banish God from the universe. If the mechanism works, if it is proven valid and sufficient, then it renders God simply redundant in that context. But that is not to say that it disproves His existence; it merely makes it a teensy bit less necessary for explanatory purposes.

I would drop the "teensy bit less necessary" business, and simply admit that if Darwin was right, then design is unnecessary for explaining the complexity of living systems. This clearly is an implication. Note that it is not an entailment. Note also that the concerns raised by this implication are squarely epistemological, not sociological. The implication states that a certain type of explanation becomes insupportable if neo-Darwinism happens to be correct, namely, any explanation that explains living systems as the product of design.

Now I agree wholeheartedly that this implication is correct. Swinburne endorses it as well. I quoted Swinburne earlier as supporting the fundamental thesis of neo-Darwinism, viz., that "complex animals and plants can be produced through generation by less complex animals and plants . . . and simple animals and plants can be produced by natural processes from inorganic matter." Swinburne makes this claim at the same time he is advancing an argument from design. How can he do this? By looking to cosmology instead of biology. Indeed, he admits that Darwin has banished design from biology.

Now Shapiro doesn't think that the implication "if neo-Darwinism is right, then design is an unnecessary explanatory device" has much riding on it theologically. For Shapiro it is enough that the existence of God remain secure. As Shapiro has rightly observed, Darwinism entails nothing about the existence of God. For theology, however, there is more at stake than simply the existence of God. The nature of this God, his relation to the world, and his causal powers to affect the world are part and parcel of any theological position.

In terms of the logic of entailment it makes no big difference to the existence of God whether Darwin was right or wrong. But in terms of the logic of implication it can make a big difference to a theological position whether Darwin was right. Shapiro's theology is certainly at peace with Darwin. A strictly pietistic theology can without much difficulty make peace with Darwin. A deistic theology can readily make peace with Darwin. Only a theology so obtuse as to insist on biblical inerrancy cannot make peace with Darwin; at least this is the impression Shapiro leaves.

What are the implications of Darwinism for theology? Shapiro has correctly argued that Darwinism does not entail any of the isms that contend with religion. Shapiro has also argued, again correctly, that Darwinism implicates the redundancy of design. Phillip Johnson (I believe rightly) takes this implication a step further, viz., Darwinism implicates naturalism. As Johnson puts it,

> "Evolution" contradicts "creation" only when it is explicitly or tacitly defined as *fully naturalistic evolution*—meaning evolution that is not directed by any purposeful intelligence.

Once one realizes that natural selection is precisely the vehicle needed to transform a theory of evolution into a *fully naturalistic* theory of evolution, the implication follows at once. Darwinism does implicate naturalism. The less God has to do, the less reason there is to maintain a theology, the more reason there is to adopt naturalism. This is not a sociological point. This is an epistemological point about the nature of explanation—about not postulating entities that are redundant or irrelevant. X therefore does implicate Y.

In closing this response, I feel it necessary to say a few words in defense of Phillip Johnson. Shapiro has charged Johnson with "making the same vulgar errors as appear routinely in the presentations of professional creationists." There is only one error I can see Shapiro referring to, and that is the error of claiming that Darwinism entails naturalism (a claim that is false simply because God can always be maintained as a useless appendage in any world view). That Johnson never claimed such an entailment should have been obvious to Shapiro, since Johnson as a lawyer is in the business of weighing evidence subject to uncertainties, and not in the business of entailments involving necessary connections. Shapiro's charge therefore cannot be supported.

Shapiro's criticism of Johnson, however, fails in a more serious way. Shapiro claims that Darwinism does not entail naturalism. Johnson claims that Darwinism implicates naturalism. Both are right. Nevertheless, I would claim that Johnson's concern in writing *Darwin on Trial* was not primarily with X entailing or implicating Y, where X is Darwinism and Y naturalism, but Y implicating X. The reverse implication is really the important one. Sure, Darwinism gives God less to do and therefore implicates naturalism. But naturalism in turn needs something like Darwinism to keep it viable.

As Alvin Plantinga puts it, if you accept naturalism, Darwinism is the only game in town. Plantinga claims an implication from naturalism to Darwinism. Johnson's work properly speaking is devoted to this implication. As a lawyer concerned with how ideological agendas—and naturalism is one such agenda—influence the courts, it is only natural for Johnson to concentrate on this implication. Shapiro has therefore missed the boat twice. The question was never whether X entails Y. It was always obvious that X implicates Y. The central question was how Y implicates X, i.e., how naturalism manages to keep Darwinism afloat. Indeed, Darwinism needs more than scientific facts to keep it afloat.

NOTE

[1]Richard Swinburne, *The Existence of God* (Oxford: Oxford University Press, 1979), p. 135.

11b
Reply to William A. Dembski
X and Y and Bob and Al and
Ted and Carol and Alice
Arthur M. Shapiro

NOT MENTIONING SOMETHING does not necessarily entail ignorance of it. Nor does it necessarily *imply* ignorance of it. Dembski is correct; entailment is the strongest logical connection. If I say that all Shapiros are geeks, this claim means that any given Shapiro is a geek. Life is comfortingly simple.

Contrast this with the claim that Shapiros *tend to be* geeks; there is some unspecified degree of connectedness between the property of Shapironess and that of geekiness. Now, suppose your daughter announces that she intends to marry a Shapiro. Are you justified in forbidding such an act, sight unseen? This involves a judgment on your part: how important is your daughter's happiness? your aversion to geeks? Do Shapiros and/or geeks have any rights that might conflict with your perceived interest? If you are a decent humane sort, rather than a flaming bigot, you would conclude that no probabilistic statement short of absolute certainty would suffice; you would insist that the putative geek be brought home for inspection. You might be less principled on this point if the matter at issue were, say, buying a post-hole digger from a Shapiro—if the price were good.

I did not discuss implication, precisely because it is so fuzzy and because the weight to be assigned it is so dependent on context. To say that methodological materialism or naturalism implies (or implicates) metaphysical materialism or naturalism is to say very little. To conflate logically distinct terms (science and scientism, evolution and evolutionism) is indeed to commit a "vulgar error," one that creates a rhetorical illusion of entailment when in fact only an unspecified, but certainly considerably weaker, association can rightfully be claimed. The more important the issue, the more inexcusable the error. (The fact that we are here argues that this issue is non-trivial.)

Phillip Johnson is a lawyer and as such is "in the business of weighing evidence subject to uncertainties, and not in the business of

175

entailments involving necessary connections." Could Dr. Dembski possibly be such a babe in the woods when it comes to lawyering? In adversarial proceedings (and if there were none such, who would need lawyers?), the lawyers "weigh evidence subject to uncertainties" in the sense that they attempt to manipulate the perceptions of others so as to minimize the appearance of uncertainty when favorable evidence for their cause is at issue, and to maximize the appearance of uncertainty when contrary evidence is at issue. That is, they attempt to create illusions of entailment or near-entailment in the minds of those "others." What "others?" Why, those who "weigh evidence subject to uncertainties" *in order to reach a judgment*, that is, judges and juries.

As a lawyer and a good one, Phillip Johnson's job—and he knows it very well—is to use rhetoric to disguise the weakness and/or unoperationality of his own claims. That's why it's important to demonstrate that the *illusion* of entailment cannot be taken for *true* entailment—because there isn't any.

After all this obscurantistic Dembskian scrapple, the last paragraph of his critique is refreshingly interesting. In it, he inverts the sense of the quote that brought us together. Remember? It says "Darwinism and neo-Darwinism . . . carry with them an *a priori* commitment to metaphysical naturalism, which is essential to make a convincing case in their behalf." But Dembski says "naturalism needs something like Darwinism to keep it viable," and therefore I have missed the boat. No, Phillip Johnson missed the boat. Dembski might be able to write an interesting paper based on this novel thesis, and I hope he does—but he had better justify his logic, because logical propositions are not automatically symmetrical, like redox reactions.

Oh yes, Al and Bob. Only once, in the second paragraph of the Al and Bob excursus, does Dembski actually say that Bob *killed* Al. As it happens, he didn't. (Lying in a pool of blood on the floor doesn't entail being dead.) The actual *dénouement* is much more interesting. Al survived, and told the police the whole story, including who shot him.

God did it. But not to worry: Phillip Johnson for the defense got him off. Charles Darwin, who wasn't even there, got forty-six years for attempted murder, aggravated assault, and naturalism in the third degree.

12
Doubts About Darwinism
Peter van Inwagen

AT THIS SYMPOSIUM we have been asked to speak on the following thesis:

> Darwinism and neo-Darwinism as generally held and taught in our society carry with them an *a priori* commitment to metaphysical naturalism, which is essential to make a convincing case on their behalf.

In order to have a label for them, I will call these words "the Quotation." I have thought about the Quotation, and I have decided that I cannot assent to it—although I by no means reject it. I have two reasons. First, I don't fully understand it, and, second, however it is to be interpreted, it is clear to me that I am not in a position to make judgments about it, owing to sheer factual ignorance.

I will take up the second point first. My ignorance pertains to the words "as generally held and taught in our society." I am not a sociologist of science, or of education, and I don't claim to know how any particular doctrine or theory is generally held and taught in our society. I admit that I've seen lot of individual bits of evidence, such as a cell biologist's quotation of a letter he was sent by a publisher telling him that a proposed textbook chapter on the origin of life should make it clear that "God is an unnecessary hypothesis," but I think I'll leave this aspect of the question alone.

As to my failure fully to understand the Quotation, this has mainly to do with the fact that different people use words in different ways, and I am not sure how the terms *Darwinism* (much less neo-*Darwinism*), *metaphysical naturalism*, and *a priori* are understood by the author of the Quotation. Each of those terms could mean more than one thing, and I know from experience that precision of meaning is important in questions about what carries commitment to what.

Although I cannot assent to the Quotation, it does not arouse in me any intellectual revulsion, but rather a sense of intellectual sympathy, a feeling that if I were to explain what I believed about the matters it touches on, someone who unreservedly agreed with it might well conclude that he and I were on essentially the same side, even if I were regrettably obtuse about several important issues. Those whose visceral reaction to the Quotation is revulsion would probably feel that I was essentially on the other side, one of the enemy. Let me explain what I believe about these matters and why I think that what I believe is true, and we shall see.

What I am going to say is perilously close to autobiography. I am not an expert on anything having to do with Darwinism. I am not even a well-informed amateur. I am just an intelligent guy who has read a few books—a very dangerous type in the world of science and scholarship, as we all know. (On the other hand, I have to point out that we are not talking about superstring theory here. The issues involved don't seem to me to be all that difficult to grasp—which is what intelligent guys who have read a few books always think.)

If you are an expert, there is only one reason you might want to pay attention to what I have to say. I am your public. If you are an expert who doesn't care what the public thinks about evolution and related matters, then you can stop reading right now. But, some experts do care what the public thinks about these things. If you are one of them, and if you think I'm wrong, I can at least tell you what it would take to convince me that you're right and I'm wrong. Then you can write me off as unteachable, or try to show me that *other* things than those I have mentioned should convince me that you're right and I'm wrong. Or you can try to do the things that I have said would convince me, or whatever takes your fancy.

I'll start by explaining how I understand "Darwinism," which seems to be the key term in these discussions. Darwinism is a theory about evolution, so I'll explain how I understand the term *evolution*. Rather than try to mark out a certain process or phenomenon that I propose to designate by this name, I'll present a series of propositions I shall describe as together constituting the thesis that evolution occurs or has occurred or is real or whatever predicate believers in evolution should use. I won't be too particular about which processes referred to in these propositions are the ones that make up the phenomenon called evolution. Since I confine the scope of my remarks to our planet, some may prefer to call my discussion "the thesis of *terrestrial* evolution." Here are the first two propositions:

- Any two living organisms, past or present, have a common ancestor.

- There have been living organisms for a very long time, not just for a few thousand years but for *millions* of thousands of years—perhaps since a few hundreds of millions of years after the earth's surface was cool enough to support life.

These two propositions taken together make up a rather weak thesis. For one thing, it is weak because it says nothing about biological diversity. This thesis could be true even if the only organisms there had ever been were a particular sort of bacterium that had persisted unchanged for billions of years. This thesis is weak also because it says almost nothing about causation—although "ancestor" is a causal concept. It is compatible, for example, with the statement that God has been responsible for a vast array of miraculous innovations in the history of life. It is also compatible with the statement that intelligent extraterrestrials have been dropping in on the earth every ten million years or so to perform prodigies of genetic engineering in aid of some mysterious agenda involving terrestrial life. To get a more interesting thesis to associate with the word *evolution*, let us add some propositions about diversity and causation.

- Life exhibits (and has exhibited for a very long time) enormous taxonomic diversity.

- Only natural causes have been at work in the production of all this diversity.

What does *natural* mean? Well, the word can be opposed both to *miraculous* or *supernatural* on the one hand, and to *artificial* on the other. Let us understand *natural* in this context as carrying both implications. The thesis of evolution implies that only the laws of physics (operating of course under an enormously complex set of boundary conditions) have been at work in the terrestrial biosphere during the course of the diversification of life. It also implies that the only extraterrestrial influences on terrestrial life have been things that are in no way the instruments of intelligence or purpose: light from the sun, cosmic rays, falls of meteor dust, asteroid strikes, and the like.

I think it is useful to regard these four propositions as together constituting the thesis of evolution. (Should there be something here suggestive of the notion of "progress," or, at any rate, of increasing complexity? Anyone who thinks so may add a clause to the effect that, in the very long run, the complexity of both the biosphere and of the most complex organisms in the biosphere tends to increase. I would not object to the addition. This seems to be a part of what a lot of people mean by evolution, and it seems to be true.)

I take Darwinism to be an identification of the "natural causes" referred to in the last of the four propositions. I take Darwinism to be a specification of a mechanism, a single mechanism, that explains taxonomic diversification. This mechanism is the operation of natural selection on random small hereditable variations that come about in the course of reproduction.

I am not, in a paper of this scope, going to try to give an exposition of what lies behind the slogan "the operation of natural selection on random small hereditable variations." I know that there is considerable diversity of opinion among those who describe themselves as Darwinians as to how the reality behind the slogan should be spelled out in detail, but I don't think that these disagreements have much to do with what I want to say. At any rate, I take it that we all have some idea of what these words mean. Even the slogan is too cumbersome for frequent repetition, so I'll call the mechanism simply "natural selection."

Darwinism, then, is the thesis of evolution plus the further thesis that the sole mechanism behind the enormous taxonomic diversity displayed by terrestrial life—behind the existence of all of those vastly different phyla and orders and classes—is natural selection. (I am aware that Darwin was probably not a Darwinian in this sense, and I am aware that he sometimes opposed natural selection to sexual selection. As to the former point, I am trying to capture at least something close to the most usual sense the word *Darwinism* has in current debates. As to the latter point, unless I am mistaken, most people today use the term *natural selection* in such a way that what Darwin called sexual selection is a special case of natural selection.)

Now where do *I* stand on all of this?

First, I accept the thesis of evolution. More exactly, I accept evolution with the exception of our own species, and even in that one very special case, I don't rule it out but merely suspend judgment. But I don't want to talk about humanity, which *is* a very special case. As a *general* thesis about taxonomic diversity, I accept the evolutionary thesis.

For example, I accept the thesis that my cat and the spider she is playing with have a common ancestor. For that matter, I believe that my cat and the spider and I have a common ancestor. To make a long story short, this seems to be the best explanation of apparently arbitrary features we have in common: the pentadactyl limb structure that the cat and I share and the genetic code that all three of us share with the algae and yeasts. I don't mean to imply that the "shared arbitrary features" argument is the only good argument for the common ancestry thesis. And I don't doubt that the lines of descent from their common ancestor to my cat and the spider involved only natural causes. To make a long story short, I believe this because I make it a rule to believe that an event or process has natural causes unless there is some reason to think otherwise, and, in the case of my cat and the spider, there seems to be no reason to think otherwise.

I accept the thesis that natural selection is *one* of the mechanisms connected with the existence of biological diversity. It has certainly been demonstrated that natural selection is a real phenomenon, a mechanism that actually operates in nature, and I see no reason to doubt that it is at least *among* the causal "inputs" that have produced the diversity of terrestrial life.

I accept the thesis that Darwinism is a genuine empirical hypothesis, and not a tautology. It is certainly true that there have been attempts to formulate Darwinism that look a lot like "in the long run, organisms that have the capacity for having the most descendants will probably have the most descendants," but I take these attempts to be simply failed attempts at formulating Darwinism. Whatever else Darwinism may imply, it implies that natural selection has—"all by itself," so to speak, without help from other mechanisms or miracles or intelligent extraterrestrials—produced enormous taxonomic diversity, and has done so within a certain measurable span of time. Darwinism therefore implies that natural selection is *capable* of doing that sort of thing, and of doing it "all by itself."

This fact suggests a thought-experiment. Suppose that we seed the oceans of millions of planets that are lifeless but suitable for life with artificial prokaryotic organisms. Suppose that these organisms have no features that would make for taxonomic diversity among their descendants other than the fact that they reproduce themselves with random small hereditable variations. (We know this because we have made them to have just that feature.) I believe that Darwinism predicts that on at least a significant proportion of these planets, we shall eventually observe biological diversity comparable with that of the present-day terrestrial biosphere: cells with nuclei, photosynthesis, multicellular organisms, sexual dimorphism, many phyla, and so on. Or perhaps we shall observe other kinds of diversity, equally striking, but without terrestrial analogue. ("Eventually"? Well, if the experiment proceeded without result for half the main-sequence lifetime of a type G star, it would then be reasonable for the granting agency to refuse further funding.)

This thought-experiment cannot be performed, but its conceivability shows that Darwinism is not in any sense a tautology, since the predicted result does not follow from the meaning of "natural selection" or the meanings of any other words: it is perfectly possible to imagine the experiment failing. I note in passing that its failure would not *refute* Darwinism—I agree with the common view that no experiment can conclusively refute a theory—but it would certainly imply that the Darwinians had some explaining to do, and that is just the kind of leverage that experimental results are supposed to have in relation to theories with genuine empirical content.

Darwinism clearly makes this prediction, and there is certainly no evidence that this prediction is not right. But it *seems* to make others, and there is evidence that some of those are not right. Darwinism seems to predict that the history of life will look a certain way: there will be few if any sharp "breaks" in that history (perhaps a few sudden extinctions of geographically confined species or genuses).

To give some intuitive sense to this prediction, suppose that we could see, laid out on a long strip of paper, a detailed picture of the father of a certain elephant, and the father of that elephant, and the father of *that* elephant, and so on. The "absence of sharp breaks" means that over millions of generations, we should see only very gradual change. A million generations ago, the animal depicted on the strip would not look very much like an elephant, but any hundred-generation section of the strip would contain only animals that looked very, very similar. And, of course, this point is intended to apply not only to elephants but to the members of any species or genus. The point applies also not to species and genuses but to any taxon: A long enough strip that starts its backward journey with a picture of a snake will somewhere contain a picture of a fish, although any hundred-generation section of the strip will contain pictures of only very similar animals.

We do not have the strip. But we do have the inevitably much less satisfactory fossil record, and it is well known that this record does not show species gradually, almost imperceptibly, shading into others as our gaze extends backward in time. As regards the broader taxa, we do not observe any line of descent that starts with, say, certain fish, and ends among the first amphibians, the members of this line becoming less and less fishy with the passage of time and acquiring more and more of the characteristics of amphibians, the intermediate members of the line being neither fish nor frog nor good red herring. Rather, we see sharp discontinuities—sharp at least as sharpness is measured on the geological time-scale, for what looks like a sudden discontinuity in the fossil record could well encompass many thousands of successive generations of organisms.

It is also well known that Darwin was troubled by the apparent discontinuities and lack of intermediate forms in the fossil record. Since Darwin's day these features have not disappeared in the light of new fossil discoveries but have become more and more evident.

On the surface, then, it looks as if Darwinism makes wrong predictions about the fossil record. But, as is usual in cases of an attractive scientific hypothesis that appears to be in conflict with some body of evidence, it is possible to devise "auxiliary hypotheses" that explain the apparent incompatibility. This has been done, if by no one else, by Gould and Eldredge, with their hypothesis that diversification takes place very rapidly among populations of peripheral isolates. As is usual in such cases, many scientists have insisted that this was just what everyone had believed all along.

When such an auxiliary hypothesis is proposed, some standard questions have to be asked: Is it coherent? Is it well motivated? Does it actually succeed in saving both the theory and the phenomena? Is the sole reason for accepting it that it saves the theory and the phenomena, or does it have something else going for it? Does the theory plus the hypothesis suggest experiments or observations that are not suggested by the theory itself?

Those are large questions. I am neither a biologist nor a philosopher of biology, and I am out of my depth here. But, speaking not as someone who claims to know anything but just as a member of the interested public, I have to say that I have not been convinced by the attempts I know of to answer them. I suppose that the main reason I am not convinced is that I am not convinced that the required intermediates are, in all cases anyway, anatomically and physiologically possible. I am not sure that a true amphibian, say, *could* be descended from a true fish across a few score thousands of generations by the small steps that Darwinism allows. I am not sure that you could take a particular fish and make a few changes in its genotype and then a few more changes and then a few more changes, and, after a few score thousand of such small sets of changes, end up with the genotype of an amphibian—not if each intermediate genotype has to be the genotype of a viable organism, and not if "a few changes" means changes of the magnitude that typically separate an organism and one of its offspring.

Let's call what I'm skeptical about the existence of "short paths": short, baby-step genetic paths between organisms belonging to, say, different biological classes. I am also skeptical about how *many* short paths exist as abstract possibilities, given that any do at all—since even if there were short paths, there might be so few of them that it would be vastly improbable that any of them would actually get taken.

Presumably, since most biologists are Darwinians of some stripe, most biologists believe that short paths exist and are numerous enough as abstract possibilities that it is not at all surprising that quite a few would actually be taken. What I should like to know more about is this: Is this belief of theirs grounded in their nuts-and-bolts knowledge of anatomy, physiology, and molecular biology? Or is it grounded simply in the fact that its truth is required by Darwinism? Unless there is some reason to believe in the existence of short paths that is prior to and independent of Darwinism, I am going to continue to be skeptical about Darwinism.

Let me recall two well-known episodes from the history of science. Newton believed that interplanetary gravitational forces rendered the solar system unstable, that, owing to cumulative distortions of the orbits of the planets by the gravitational fields of the other planets, the solar system could not retain its dynamic stability for more than a few centuries. He dealt with this difficulty by postulating periodic divine corrections of the planetary orbits. To remove a red herring, let us pretend that he postulated not miraculous interference in the course of nature, but rather the action of some as yet unknown physical principle, in addition to the laws of motion and the law of universal gravitation. A generation or so after Newton, Laplace showed that the destabilizing effects of mutual planetary gravitational attraction that Newton worried about tended to cancel out, and that, although a solar system whose motions were governed solely by the laws of motion and gravitation was perhaps not *absolutely* stable, it would be capable of retaining its stability over vast stretches of time.

Lord Kelvin insisted that, despite what the paleontologists said, the sun could not have been shining at its present luminosity for more than a score or so millions of years. This was because that is the longest period you could get on any reasonable initial conditions if solar radiation was, as he supposed, due entirely to the release of gravitational potential energy in the form of radiation as the material of the sun underwent gravitational contraction.

In my view, owing to the difficulties I have briefly mentioned, Darwinism is in the position either of classical celestial mechanics in the time of Newton, or else in the position of the standard late-nineteenth century theory of solar radiation that Kelvin appealed to. In each case the theory *appears* to make the wrong predictions about the observed phenomenon. Newton knew it. Kelvin denied it, dismissing the claims of paleontology as confidently as any twentieth century "creation scientist." In the case of Newton and Laplace, the difficulty was surmountable, although surmounting it was by no means trivial. It required all the resources of one of the greatest applied mathematicians in history. In the latter case, the difficulty was insurmountable. Kelvin's proposed mechanism (the transformation of gravitational potential energy to radiant energy) is *there* all right, but it is one of several mechanisms that contribute to solar radiation, and the others are responsible for the lion's share of the effect. Lord Kelvin's implicit theory, that only the one mechanism was at work, was wrong.

Which of these cases represents the situation of Darwinism? Well, I am inclined to think the second. Those who say that there is no problem are in roughly the position of Lord Kelvin vis-à-vis the data of paleontology. If the situation of Darwinism is analogous to the first case, we do not now know this. In that event, evolution has had in Darwin its Newton, but it has not yet had its Laplace. If the situation of Darwinism is analogous to the second case, then there are as yet undiscovered evolutionary mechanisms, ones that contribute the lion's share of the effect. (I should mention that the analogies I have been appealing to have at least one serious defect. Classical gravitational mechanics is a quantitative theory, and it is pretty clear what its predictions are. It is not the fault of Darwinism that it is not a quantitative theory, but the fact that it is not does have the consequence that it is much less clear what its predictions are.)

I am not quite finished with the case of Lord Kelvin. Before leaving it, I want to use it as a stick with which to beat the following argument: "No one should say that evolution requires other mechanisms than natural selection unless he or she has some constructive proposal to

185

make about what those mechanisms might be." I have heard somewhere that, as a matter of fact, some paleontologists did rather timidly ask Kelvin whether there might be some unknown factor involved in the production of solar radiation. His reply was evidently contemptuous and dismissive. He might well have used an argument exactly parallel to the one we are considering: You shouldn't make that suggestion unless you have some constructive proposal to make about what that factor might be. If he had said this, he would have been wrong. He should have been willing to admit that paleontological evidence, in conjunction with his own calculations, established at least a very strong prima facie case for the conclusion that some factor other than gravitational contraction was partly responsible for the sun's energy output. He should have been willing to admit this despite the fact that no physicist, and certainly no paleontologist, had any constructive suggestion about what that factor might be. (We know now that any speculation about this question at the turn of the twentieth century would have been a waste of time.)

So that is where I stand. It looks to me as if natural selection is not a complete explanation of the diversity of life. I am inclined to think that its primary "function," if I may use that word, is to insure that species possess sufficient diachronic flexibility that they aren't just automatically wiped out by the first environmental change that comes along. And, of course, natural selection is a very efficient fine-tuning mechanism: once a species has found an ecological niche for itself, natural selection tends to optimize its "fit" into that niche.

And I am willing to allow a little more to natural selection than this. I am inclined to think that "unaided" natural selection *can* produce new species; I have a very hard time believing that it can produce, say, new classes. There are (or so it looks from where I stand—not much of a vantage-point, I admit) mechanisms involved in biological diversification that are as unknown, and probably as unguessable, today as the release of surplus binding energy in nuclear fusion was in the year 1900. (But I don't mean to suggest that these mechanisms involve new physical principles.) It looks to me as if Darwinians are like someone who, having observed that tugboats sometimes maneuver ocean liners in tight places by directing high-pressure streams of water at them, concludes that he has discovered the method by which the liners cross the Atlantic.

Now a concluding even more unscientific postscript, connecting what I have said so far with my religious views. Like St. Augustine, I am not a literalist about the first three chapters of Genesis. Writing early in the fifth century, Augustine held that the six "days" of creation in Genesis were not meant to be taken as literal twenty-four hour days, but were a rhetorical figure used to describe six aspects of creation. He held that in the beginning the world contained much less actual order than it does today, and that the order we now observe in the world evolved—that is, "unfolded"—out of the potential order that God had placed in things at the moment of creation. This would be my view as well. I see it as the business of science to uncover the mechanisms of that unfolding.

As to biological order, if unaided natural selection really is capable of producing the ordered diversity we see in the terrestrial biosphere today, I see no reason why a God who wanted such ordered diversity should not have used this very elegant mechanism. If I doubt that God did this, it is only because I doubt that unaided natural selection could do the job. I think that other mechanisms would be required and that he therefore must have used them. But if unaided natural selection would work—well, why shouldn't God use something that would work?

It seems to be a widespread opinion that something about natural selection unfits it for use as a divine instrument. I have never been able to see this. When I was an agnostic, I was a Darwinian. When I became a Christian, a very old-fashioned, orthodox one, I was a Darwinian still. And although I have experienced many intellectual difficulties with my faith, my belief in Darwinism never caused me the least intellectual discomfort. My doubts about Darwinism began only when I discovered that the "smoothness" of the fossil record that I had always believed in was not there. I should add, in this connection, that I do not regard the difficulties that I believe Darwinism faces as constituting any sort of evidence of theism. I think that the truth or falsity of Darwinism has no more to do with theism than does, say, the hypothesis of continental drift.

But many people do not see things this way. I could quote both Darwinians and anti-Darwinians to this effect. Here is a famous quotation from Monod that will do as well as any. Speaking of the events that have been identified as the sources of mutations, he says:

> We call these events accidental; we say that they are random occurrences. And since they constitute the *only* possible source of modifications in the genetic text, itself the *sole* repository of the organism's hereditary structure, it necessarily follows that chance alone is at the source of every innovation, of all creation

187

in the biosphere. Pure chance, absolutely free but blind, at the very root of the stupendous edifice of evolution: this central concept of modern biology . . . is today the *sole* conceivable hypothesis, the only one that squares with observed and tested fact.[1]

Monod goes on to make clear that he understands chance in Aristotle's sense, as arising from the coincidence of independent lines of causation. (Thus, it is due to chance that Shakespeare and Cervantes died on the same day, as it would not be if they had killed each other in a duel. In this sense, chance can exist even in a fully deterministic world.) He identifies the source of this chance with imperfections in the fundamental mechanisms of molecular invariance in living organisms. He mentions only the causes of mutations, but he might have mentioned other sorts of events that are of evolutionary significance and can with equal plausibility be ascribed to chance: the flood that happened to destroy a certain herd of ruminants, the raising by geological forces of a land bridge that enabled representatives of certain species to move into a new environment, the intersection of the trajectories of the earth and a certain comet, and so on.

I don't quite see how it is that the hypothesis that all such events are due to chance is the only conceivable hypothesis. But let us suppose that this hypothesis is at any rate *true*. Does it follow that the general features of the biosphere are products of chance? It does not. To suppose that they are would be to commit the so-called fallacy of composition. It would be as if one reasoned that because a cow is entirely composed of quarks and electrons, and quarks and electrons are nonliving and invisible, a cow must therefore be nonliving and invisible.

There is a marvelous device for calculating the areas surrounded by irregular closed curves. It is an electronic realization of what is sometimes called the dartboard technique. To simplify somewhat: you draw the curve on a screen; then the device selects points on the screen at random, and looks to see whether or not each point falls inside the curve; as the number of points chosen increases, the ratio of the chosen points that fall inside the curve to the total number of chosen points tends to the ratio of the area enclosed by the curve to the area of the screen. For a large class of curves, including all that you could draw by hand, and probably all that would be of practical interest to scientists or engineers, the convergence of ratios is quite rapid. Because of this, such devices are useful and have been built. Now the properties of each point that is chosen, its coordinates, are products of chance in just Monod's sense. But the whole assemblage of points chosen in the course of solving a given area problem has an important property that is not due to chance: its capacity to represent the area of a curve that had been drawn before any of the points was chosen.

Indeed, since the device was built by purposive beings, there can be no objection to saying that the whole assemblage of points has the *purpose* of representing the area of that curve—despite the fact that the coordinates of each individual point have no purpose whatever. It is also true that the fact that each point has coordinates that are due to chance is not due to chance and has a purpose: its purpose is the elimination of bias, to insure that the probability of a given point's falling inside the curve depends on the proportion of the screen enclosed by the curve and on nothing else.

Suppose that every mutation that has ever occurred is, as Monod says, due to chance. Suppose, in fact, that every individual event of any kind that is a part of the causal history of the biosphere is due to chance. It does not follow that every aspect of the biosphere is due to chance. And if none of these individual events has a purpose, it does not follow that the biosphere has no purpose. To make either inference is to commit the fallacy of composition.

Now this reasoning shows at most that the thesis that some features of the biosphere are not due to chance (and likewise the stronger thesis that they have a purpose) is logically consistent with Darwinism. It could still be that the conditional probability of the thesis that there are features of the biosphere that are not due to chance is very low, even negligible, on the hypothesis of Darwinism. But the reasoning does show that if someone wants to construct an *argument* for the conclusion that Darwinism is in any sense incompatible with the thesis that some features of the biosphere are not products of chance, he or she will have to employ some premise in addition to "Darwinism implies that all events of evolutionary significance are due to chance." And, as I have implied, I do not find that premise itself indisputable.

One argument might be that the features of the biosphere are in a very important respect unlike the features of an assemblage of points produced by our area-measuring device. Each time we draw a curve on the screen of the area-measurer and turn the thing on, it is for all practical purposes determined, foreordained, that the assemblage of points it produces will have the property of representing the area enclosed by the curve.

But, it might be argued, the properties of the biosphere are not like that. There used to be a popular thesis called Biochemical Predestination, according to which they *were* like that. According to Biochemical Predestination, you just take a lifeless planet that satisfies certain conditions (conditions the earth satisfied before there was any life on it, and which are undemanding enough that it would be reasonable to suppose that a pretty fair number of planets in a given galaxy satisfied them) and in due course you will "automatically" have life, eukaryotic life, multicellular life, sexually dimorphic life, highly differentiated life, and, finally, intelligent life—the whole *Star Trek* scenario.

Biochemical Predestination does not seem to be very popular among the practitioners of the life sciences these days, although belief in it seems to be common among physicists and astronomers and nearly universal among university undergraduates, who believe that Vulcans and Klingons await us among the stars with the same unreflective assurance that attended the belief of their twenty-times-great grandparents that elves and trolls awaited them in the woods. But if Biochemical Predestination is not true, if the main features of the biosphere did not fall into place automatically, but rather are due to remote chances that just happened to come off, then how can it be that these features are due to the purposes of a divine being—or any intelligent being? In short, the failure of Biochemical Predestination shows that, since the evolutionary process has no determinate "output," it is not the kind of thing that could be anyone's instrument.

Curiously enough, Biochemical Predestination was said by those who believed in it to show that the evolutionary process was not anyone's instrument, owing to the fact that, according to that hypothesis, the features of the biosphere are a consequence of the laws of physics operating on the matter near the surface of the earth, and have therefore been produced without any need for manipulation by outside forces. Moreover, since these same features would have emerged from almost any set of initial conditions, they have been produced without any need for any sort of initial adjustment or fine-tuning of the state of the matter near the surface of the earth.

I don't myself see the force of either of these ideas. I don't see why either Biochemical Predestination or its denial should be thought to have any theological (or atheological) implications. Perhaps what is needed in order for there to be a useful discussion of the question whether there are such implications is some measure of agreement about what a biosphere that was a divine creation would look like: what it would look like at any given point in time, and what its history would look like. After all, if you propose to refute an hypothesis by an appeal to observation, you have to have some idea about what things would look like if that hypothesis were true.

I myself have almost no expectations about what a divinely created biosphere would look like. I mean I have no *a priori* expectations. Since I think that the biosphere is in fact a divine creation, I of course think I know *one* thing a divinely created biosphere might look like: what it does look like. How should *I* know what features to expect a biosphere to have if that biosphere were created by a being whose knowledge and wisdom were unlimited and whose power was limited only by considerations of what is intrinsically possible? Before I could make even a guess, I should have to know what that being wanted the biosphere *for*, and I should have to know a lot more than I do about what is intrinsically possible. I don't see how anyone could know what a divine being wanted a biosphere for—not unless the divine being told him, anyway. And I doubt whether anyone knows much more than I do—much more than almost nothing at all—about what is intrinsically possible.

NOTE

[1]Jacques Monod, *Chance and Necessity: An Essay on the Natural Philosophy of Modern Biology*, tr. Austryn Wainhouse (New York: Vintage Books, 1971), pp. 112-113.

12a
Response to Peter van Inwagen
The Problem of Language
Frederick Grinnell

"THE FUNDAMENTALISTS," WROTE Jewish philosopher Abraham Heschel, "claim that all ultimate questions have been answered; the logical positivists maintain that all ultimate questions are meaningless."[1] Professor van Inwagen and I are somewhere in between, concerned about the questions, but not sure of all the answers.

In his paper, Professor van Inwagen presents a kind of systematic doubt grounded in language. He doesn't understand precisely what Darwinism means or what metaphysical naturalism means, or at least he realizes that these terms may have quite different meanings depending on the user. Also, he cannot be sure what a divinely created biosphere would look like in an *a priori* sense without knowing the purpose for which the biosphere was created. Consequently, it is difficult to draw clear relationships between Darwinism, metaphysical naturalism, and the possibility of a divinely created biosphere.

His focus on the problem of language, on the problem of what particular words mean to the persons who use them, is a key point in any discussion about Darwinism or about science in general. In my response, I want to emphasize and elaborate on this point.

In all social interactions, we communicate with each other according to typical expectations of what sort of language would be appropriate. Some of us who share common interests and activities (for instance, religious or scientific) use language in group-specific ways. When particular words have widely different meanings according to the background and expectations of the speaker and listener, the possibility for confusion increases markedly.

Language confounds our discussion about Darwinism at two levels: the first at the level of communication, the second at the level of imagination. When I read the statement, "Darwinism and neo-Darwinism as generally held and taught in our society carry with them an *a priori* commitment to metaphysical naturalism," it seems backwards to me. If I wanted to link Darwinism and metaphysical naturalism, I would have written the following. "Metaphysical naturalism is an *a priori* assumption that makes doing science possible, which includes Darwinism and neo-Darwinism."

As I have described in more detail in my chapter, the assumption of naturalism is necessary for scientists in order to make their research credible. Only by assuming that their research can be verified by others can individual scientists transcend their subjectivity. It is precisely this assumption that grounds the objectivity of science. That is, I assume that my experimental results are not an outcome of my personal biases since I believe that they can be seen and verified, at least potentially, by everyone else. In short, whatever cannot be measured or counted or photographed cannot be science, even if it is important. Therefore, when a writer implies that Darwinism could be separated from an *a priori* commitment to metaphysical naturalism, I know that he and I understand science to mean different things. We haven't shared the experience of doing science and of trying to make science credible.

The problem of different meanings of the same word can be overcome, at least in part, by trying to make explicit to each other what we mean by the words we use.

The second problem relating to language, that of imagination, is more difficult to overcome. Like other activities of daily life, science depends upon human language for its description. Paradoxically, however, many scientific concepts eventually refer to aspects of reality beyond the possibility of common experience. That is, although science begins with the language of common experience, it often produces descriptions that not only lose their direct connection to, but also may contradict, routine experience. According to quantum physics, tables are mostly empty space, but they feel solid to me. I have trouble imagining an expanding universe. Greek science must have had a tough time convincing people that the earth was spherical rather than flat.

Therefore, despite our attempts at clarity, many scientific ideas are difficult to think about because they cannot be expressed clearly using descriptive language. The situation is like trying to explain to someone who has never seen a red object what the color red looks like. Simply telling the person about the physical events involved in seeing red color—that is, light of a certain wavelength interacts with pigments in the photoreceptors of an observer's eye, etc.—misses entirely the sense of personal experience of redness.

The most obvious differences between science and everyday

experience occur in physics, which deals with objects that are very large, very small, and very fast compared to those we normally encounter. Biological thought and language, which is the focus of this conference, present a problem because the objects of biology have histories, histories that count. To describe the evolving characteristics of a group of organisms, one must learn to think in four dimensions, three dimensions of space stretched across a very long dimension of time. The only way I can even begin to imagine what evolutionary thinking might be like is to try to look at my friend as an integrated historical sequence rather than as the individual who confronts me here and now. Not an easy task.

In addition, the uniqueness of such historical sequences impedes usual scientific thinking, which does best when dealing with recurring events. Far from the reductionist ideal, evolutionary biology requires a holistic approach to science. Most people find holistic thinking difficult. In general, the move away from reductionism has about as much appeal as the uncertainty principle of quantum mechanics had for Albert Einstein. It is not scientists alone, however, who prefer reductionistic descriptions. It took New Testament scholars 1,800 years to begin a hermeneutic approach to biblical interpretation.

In summary, it does not surprise me that Professor van Inwagen has noticed the variable meanings and implications of Darwinism. Like much of science, understanding Darwinism requires us to use our imagination in novel ways that go beyond everyday experience, to use conceptual and mathematic models that can only be approximated by everyday language. That is why we argue about precisely what the models mean. That is why our understanding of Darwinism itself continues to evolve. At any stage, however, what makes different models appear credible in a scientific sense—in the way that I mean science—is their potential for verification, and this verification can occur only in the naturalistic world shared by everyone.

NOTE

[1]Heschel, A. J., *God in Search of Man: A Philosophy of Judaism*, Farrar, Straus, and Giroux Publishers, New York, 1955.

13
A Blindfolded Watchmaker:
The Arrival of the Fittest
David L. Wilcox

WHY HAS THE NEO-DARWINIAN paradigm become the accepted explanation for the biological world? This is the issue before this symposium. Has its endorsement been due to its perceived metaphysical necessity, or is it due to its success as a scientific explanation of empirical phenomena?

If I am to speak to this issue, I want to make my focus very clear. This paper concerns the appearance of biological structure, *not* the tie of such appearance to biotic descent. Evidence for structural difference/ descent does *not* constitute evidence for the mechanism by which structural transformation took place. Therefore, the sorts of evidence that simply indicate relationship and/or descent from a common ancestor (e.g., molecular clock data, fossil sequences, chromosomal banding, and other measures of similarity) are *not* relevant to this question unless they indicate the nature of the creative mechanism that produced novelty during that descent. Evidence of ancestry does *not* imply knowledge of the morphogenetic mechanisms that are able to produce novelty.

This was perhaps better understood in the nineteenth century than it is today (Müller and Wagner, 1991). Indeed, by 1850, almost *all* researchers accepted common descent (Gillespie, 1979; Desmond, 1989). The *unique* implication of Darwin's theory was therefore not descent, but its suggestion that the source of biotic order was to be found in the natural (material) order. For the "Naturalist" (Materialist) of Huxley's Young Guard, natural selection was not simply a theory of mechanism, but a replacement for the Creator (Desmond, 1989; Moore, 1982).

It still is. From the time Darwin proposed it, the central hope of neo-Darwinian theory has been its supposed ability to remove the need for and to take the place of an immaterial designer. According to Stephen Gould (1982), "Natural Selection is a creator—it builds adaptation step by step." As G. G. Simpson (1967) put it,

> It is already evident that all the objective phenomena of the

history of life can be explained by purely naturalistic, or in the proper meaning of a much abused word, materialistic factors. They are readily explicable on the basis of differential reproduction in populations (the main factor in the modern conception of natural selection) and of the mainly random interplay of the known processes of heredity. . . . Man is the result of a purposeless and natural process that did not have him in mind.

Clearly, if the biosphere is self-realizing and unguided, a designer without goals, Richard Dawkins was justified in his remark that "Darwin made it possible to be an intellectually fulfilled atheist." In that sense, Darwin's "scientific" theory forms a *necessary* support for the beliefs of the committed materialist. This does raise an immediate concern, since it is very difficult for true believers to be objective about proofs concerning the foundational assumptions of their faith.

But has the Darwinian hope proved justified? Or, have cracks in the explanatory plaster been papered over (by faith)? My thesis for this paper is that the plausibility of the neo-Darwinian hypothesis as an explanation for the appearance of biological novelty depends on an inadequately simple model of the genome.

The first problem is a matter of simple logic. How can *natural selection* be Gould's creator of new morphology when it does not write genetic messages, but only chooses between them? Rather than a creator, it is a critic. It brings no information into the genome, but only selects forms already "created" by the mutated genome. Michelangelo once said he did not carve an angel, he only released it from the rock. For the artist this was modesty, but for selection it is simple truth. The "grain" of the wood being carved, i.e., the informational characteristics of the genome itself and the probability structure of genetic phase space (Brooks et al., 1989, define GPS as the probability space of all possible genomes), determine what selection is able to produce. One cannot select a characteristic not already present; a horse breeder cannot produce Pegasus.

Clearly, then, the nature of the information encoded on the genome, the genetic programs that can be mutated, are central to understanding natural selection's ability to "create." The information structure, however, is far more complex than has been usually assumed. Specifically, the information encoded on the genome reveals hierarchy. It is a hierarchy of described reality, of blueprints for specific cell types, organs, organisms, hives—blueprints of immense stability. Further, these blueprints are organized in a form analogous to a linguistic hierarchy, in which the definitions of the "markers" are written in the code they define—e.g., amino acid code in the encoded description of the structure of the aminoacyl proteins. Nor are these simple descriptions of morphology, but a cybernetic hierarchy of controls, with the goals of the more comprehensive levels buffered by flexibility in the lower levels. Finally, those goals are realized through a temporal (developmental) hierarchy, in which the same goal may be achieved by different paths (Müller and Wagner, 1991). The information that dictates error-checked homeostatic/homeorhytic phenomena must itself be error-checked and cybernetic. Clearly, then, there are at least two classes of morphological information on the genome: adaptive and prescriptive. The implication for our topic is that the evidence for selective movement in the adaptive class does not prove that selection can move or formulate prescriptive information.

The other half of the Darwinian engine is *mutation*. Since the *environment* can select only new variants tossed up by the genome (by mutation, recombination, etc.), for morphology to be due to selective direction, mutation would have to produce uniform additive phenotypic variation. Mutation, however, seems to be an adaptive resource utilized *by* the existing hierarchical genome, rather than a simple mechanism to "broaden" the phenotype. The existing *genome* is the selective agent. Borstnik et al. (1987) report that random mutation acts as a search process, and Parsons (1987, 1988, 1991) indicates that mutational rates increase in stressed populations. Pakula and Saner (1989) base the evaluation of new mutants on the function of the original sequence, and Wilkins (1980) states that mutated genes must be understood within their higher level constraints. Belyaev (1979) demonstrated that very high levels of genetic variability are masked in fox populations, and Wilkins (1980) that assorted eye mutants all affect the size of the entire eye. Turelli (1988) reports that mutation rates per *character* are three-or-four-fold higher than the rate per *locus*.

Clearly then, mutation plays an adaptive role for some genetically coherent levels (entities). A "new" gene's action is constrained by its purpose within a more comprehensive biotic entity (its role within a higher set of rules or blueprint). The existing genome controls the meaning of new mutants. Thus, it is the "old" genome, rather than the environment, that is the matrix/source of new morphology. But will that same process *produce* those genetic entities? Changes/novelties must make sense in terms of the complete error-checked genomic system, or the mutant organism (a genetic trial balloon) will not mature and reproduce. Given a rabbit in the hat, a magician can pull it out—but how do rabbits get into the genetic hat? That, too, natural selection must answer if it is to be the genetic maestro.

Can *natural selection* provide the constraint for the genome? Models designed to explain the bases of such constraints have been problematic. Available genetic diversity is too high to be a constraint. Observed morphologies have moved as much as ten standard deviations in response to selection. Stabile environments (niche space), the ability of populations to track favorable environments, and stabilizing selection all predict low diversity. As Barton and Turelli (1989) put it, "the central paradox is that we see abundant polygenic variation, together with stabilizing selection that is expected to eliminate that variation." Direct replacement (Lande, 1975; Yoo, 1980) for the mutant alleles must be almost identical to those lost to selection (Barton and Turelli, 1989). Barton and Turelli's own stasis theory of pleiotropy predicts only short term stasis; since the dysfunctional effects are diffuse and randomized, the genes involved will be subject to slow replacement. Studies (Turelli, 1988; Wilkins, 1980) that show three-to-four-fold higher mutation rate per character than per locus clearly show the constraint of the existing genomic blueprints. And again, the laboratory/field evidence (Endler, 1986; Boag, 1981) of simple changes in gene frequency due to selection, drift, etc., are not significant unless they can be shown to be capable of leading to true novelty, that is, to coherent new morphologies rather than just to shifts in the diversity pattern of the adaptive genome.

Such a complex genome is unlikely to be changed simply by random mutation. It is no block of uniform marble to be chipped away by random hammer blows, but a series of gnarled and knotted instruction sets that must be courted and wooed if change is to be achieved. This is clearly the meaning of the evidence given above for the complexity of mutational change. But mutation is not the only area of investigation that supports a cybernetic information structure on the genome.

For instance, the nature of *species* is an open question. The best definition of species seems to be based on a cohesion model rather than on reproductive isolation. Paterson (1985) and Templeton (1989) have pointed out inadequacies in the usual species definitions of reproductive coherence or isolation. For instance, parthenogenic species may remain morphologically coherent despite a total lack of interbreeding. Or, in syngameons, interbreeding species may remain morphologically stabile and separated despite millions of years of gene flow. Templeton suggests that species' identity is due to their possession of various genetically based cohesion mechanisms; thus species are characterized by specific individuated genetic controls. Burton and Hewitt (1989) report that such mechanisms control species boundaries. Further, Vrba's (1984) work with the alcelaphine tribe of African antelope suggests that the whole tribe might be viewed as an entity controlled by such a common coherency.

Cybernetic models of coherency are also important in *developmental* biology. According to Wagner (1989), "Anatomy emerges at the level of the organ but not at the level of the parts." He refers to control by such sets of developmental constraints as "individuation" or entity formation. The organ is thus ontologically prior to the parts; it defines them and gives them a local "purpose" or limited "final cause." Bryant and Simpson (1984) also speak of "emergence" as a characteristic of a group of cells committed to form an organ, and error-checked according to norms for that structure. Thus, an adequate understanding of embryonic tissues involves their *purpose* to the forming organ, and implies the existence of a genomic organ "blueprint" (Wagner, 1989).

The tension between natural selection and developmental coherency is evident in two recent papers by Weber (1992) and Wake (1991). Weber states that ". . . very small regions of morphology (less than 100 cells across) can respond to selection almost independently . . . ." Wake states that the common phenomenon of amphibian homoplasy is due to "limited developmental and structural options," i.e., to design limitations. The power of developmental individuations (cybernetic coherencies) is shown in the fact that existing diversity is utilized to ensure morphological stasis rather than directional change.

Wagner (1989) also suggests that homology should be centered around shared entity formation. "Structures from two individuals or from the same individual are homologous if they share a set of developmental constraints, caused by locally acting self-regulatory mechanisms of organ differentiation. These structures are thus developmentally individualized parts of the phenotype." Such a view of homology would give a meaningful approach to several ongoing difficulties, including the phylogenetic reappearance of "lost" structures (e.g., avian clavicles, Bakker, 1986); alternate inducers of the same organs (Hall, 1983); alternate paths of development in related species

(Raff and Kaufmann, 1983); the use of same control genes in different developmental pathways (Marx, 1992) termed genetic piracy by Roth (1988); iterative homology (parallels in repeated organs) (Müller and Wagner, 1991), and the growth of "homologous" organs from different embryonic primordia or germ layers (Wagner, 1989).

Developmental individuation again demonstrates that the GPS is gnarled and knotted rather than uniform marble. Goodwin concludes that ". . . the organismic domain as a whole has a 'form' and is therefore, intelligible (which does not mean predictable) and that the 'content'—the diversity of living forms, or at least their essential features—can be accounted for in terms of a relatively small number of generative rules or laws" (Webster and Goodwin, 1982). Such existing blueprints constrain selection into a few possible paths. Rieppel (1990) agrees that some sort of morphogenetic "generative principles" dictates the possibilities of biological form. Such structural rules would restrict living things to parts of GPS that contain permitted morphologies. As Goodwin (Webster and Goodwin, 1982) put it, "A 'generative structuralism' is required in order to solve the problem of the origin of structures." Again, ". . . living organisms are devices which use the contingent 'noise' of history as a 'motor' to explore the set of structures, perhaps infinitely large, which are possible for them." But, how are such curious devices first formed?

Individuations as coherent genetic entities are fundamental biotic realities. But that raises questions: What is the origin of such sets of cybernetic constraints/new individuations? How effectively can natural selection produce or modify such a coherency? Why is it that the evident structure of Genetic Phase Space is so convoluted? What are the density and distribution of coherent, viable blueprints in GPS?

Hard questions. Certainly GPS, which is the probability space of all possible genomes, and thus of all possible genomic coherences, is so large as to be beyond comprehension, much less prediction. The information content of GPS is 2^n bits where n is the number of bases. The GPS of genomes of mammalian size (2.5 billion bases) contains around $10^{1,000,000,000}$ binary bits of information. In contrast, Dawkins's Biomorph land (Dawkins, 1986) has a probability space of only 10^{15}. What can we know of GPS? If Dawkins's little predefined universe contains a mysterious, unpredictable "Holy Grail," surely the probabilities of outcomes in GPS cannot be known. Nor can we demand that it have some specific probability structure *so that* neo-Darwinism will work. Or rather, we can demand that *only* if we first *assume* that neo-Darwinism works, but that competence is what we are trying to *prove*, is it not?

If direct knowledge of GPS total structure is impossible, all we have is inferences concerning its *local* structure, which can be drawn only from the pattern of the fossil record, the record of the search. There is no other evidence. But the fossil record shows an unevenness of rate suggesting coherencies, and that evidence throws doubt on the adequacy of neo-Darwinism as a creative source of new morphology. If it cannot explain, why is it accepted? I note the following problem areas.

1. *Life's origin.* The origin of life requires the initial encoding of specified blueprints, a non-Darwinian process. Specification involves arbitrary definitions for the "letters" used to write the "messages." How then did specified complexity (blueprints and their described products/"machines") arise from any amount of nonspecified complexity (complex machines, but no blueprints)? Are we really making progress in explaining the source of the genetic code? "The holy grail is to combine information content with replication" (Orgel in Amato, 1992). That is, we need a machine that can write down its own specifications (Thaxton, 1984).

2. *Origin of the first animals (Cambrian era).* The Cambrian explosion illustrates the abrupt formulation of body-plan constraints (Erwin, et al. 1987). But how within 25 million years (impalas have remained unchanged longer than that) could the full complexity of 70-plus metazoan phylum level body-plans arise, and be individuated with error-checking developmental cybernetic controls from protozoans? Remember that protozoans do *not* have encoded genetic information for morphology due to cellular interaction. How can code that does not yet exist be mutated? Further, given the appearance of new code, how are phylum level morphological "norms" generated, capable of holding for the remainder of the Phanerozoic? As David Jablonski put it, "The most dramatic kinds of evolutionary novelty, major innovations, are among the least understood components of the evolutionary process" (Lewin, 1988).

3. *Species stasis.* Species show morphological stasis in the face of high levels of selectable diversity (Stanley, 1979 & 1985). But what sort of genetic anchor can hold constant a species' morphological mean *and variance* for several million years (Michaux, 1989), when enough genetic diversity exists in such species to allow laboratory selection to

cause a ten-fold movement of that morphological mean? Are current models of the informational organization of the genome adequate to explain this? This difficulty is reinforced by the still greater morphological stasis shown by the body-plans of the higher levels of the taxonomic system, a stasis that seems to shape, direct, and constrain lower level change in an almost "archetypic" manner. This is hardly the neo-Darwinian prediction.

4. *Sudden individuation.* New individuation, the appearance of adaptive complexes (morphological entities) is typically very abrupt—for instance, limb structure in *Diacodexus* (Rose, 1982 & 1987) or the Ichthyostegeds (Coates and Clack, 1991). New "type" forms usually appear suddenly, with the characteristic morphological systems already "individuated"—as defined and error-checked entities. (Such definition will almost always require more "bytes" to encode.) Even if possible ancestors that lack the new complex seem to be present (usually at about the same point in time), where do the new control system norms come from? The appearance of new taxa seems to imply the sudden appearance of packages of individuated structural information, but how does closed, error-checked cybernetic feedback start? It may be relatively easy to show that a path across *phenotypic* space could be progressively adaptive (Kingsolver and Koehl, 1985), but explaining the necessary changes in the underlying genome is a different matter. The two seem identical only because neo-Darwinism has assumed the supply of sufficient additive variability.

The origin of individuation is not an easy question (Müller and Wagner, 1991). To make insects from centipedes, three segments must form a new individuated entity, the thorax. For that to happen, there must be a new set of constraints encoded on the genome for the thorax, rules that define the new entity. Such a rule-set requires a lot more information than did the original repeated structures (segments). The genomic change is far more complex than the phenotypic change. Wagner (1989) states that we have no way "to assess the plausibility of the internalization mechanism . . . the relevant type of data is not thus far available."

5. *Mosaic evolution at morphogenic transitions.* Intermediate evidence, when it does exist, usually is mosaic in nature. Mosaic evolution (the movement of one character with stasis in another) indicates the constraints of existing genomic diversity. But, if the characteristic appearance of new suites of characters is similar to that

seen in *Archeopteryx*, then an almost completely established (individuated) character set can be obtained for one organ/structure (flight feathers) with little movement in others (skeletal characteristics) (Wellnhofer, 1990; Sereno and Chenggang, 1992). This makes sense only if the complexity to be realized was already available in the genome. If large-scale morphological change depends on the appearance of a series of new mutations to be selected by a new adaptive niche, should not characters be mutated and move together at rates that are at least comparable?

6. *Adaptive radiations.* The speed, character, and commonness of adaptive radiations indicate the partitioning and exploration of an occasionally rich genome. Almost all groups at all taxonomic levels first appear in the record as "type" forms, and then "explode" into a number of different lineages with a mosaic of related but not identical potentials for adaptive morphological change (see #5 and the wealth of information in Carroll, 1988; MacFadden and Hulbert, 1988; Larson, 1989). This pattern suggests the partitioning of a very large common genetic package with a high number of alternate morphological potentials. But no known mechanism is available for generating such information-dense primordial genomes. Selection can act only on phenotype, not on hidden genetic potentials. The idea that a "key" innovation opens a "new" adaptive field assumes what needs to be proved about the ability of a genome to be reconfigured in multiple ways. As a matter of fact, a "key" adaptation would be more likely to produce a plethora of pleiotropic dysfunctions.

7. *Parallel development in lineages.* In adaptive radiations, the diverging lineages will frequently develop in a parallel fashion for a number of characteristics. Such parallels can be quite detailed, suggesting that distantly related species are relatively close. This implies that potentials for the parallel developments were already present in the parental genome as coherent potential blueprints. Thus, "convergent" evolution frequently looks as if it is due more to shared genomic constraints than to shared environments. To what extent can "random" mutations be expected to parallel each other?

So then, we have seen that selection does not "create" anything, but it must already be there for selection to find, and thus biological novelty must be generated by the entire genome. Further, we noted that numerous areas of biological investigation (the nature of mutation, species, development, and homology) point to a genome constructed as a hierarchy of cybernetic individuations. Finally, since the GPS is far too large to predict outcomes, the only way we have to evaluate even its local structure is the fossil record itself. The best evidence for selection appears to be the sorting of packages of existing genetic blueprints, not their creation (or location in GPS). Clearly, the GPS locale being searched by biotic lineages is extremely complex. But the mechanisms for the appearance of such novel packages (or the finding of such remote GPS locations/probabilities) remains mysterious. Thus, natural selection has *not* been shown to be an effective creator-substitute. It fails at just the point where it must succeed.

Of course, it is possible to postulate a structure for GPS that can be explored by random search processes, a structure that would (if we could see it) predict the world as it exists. In fact, there is no way to prove that GPS is *not* structured in this fashion. No empirical evidence can be raised against this possibility, because the necessary precursors of the evidence could be "programmed into" the model of GPS. It seems that a blind watchmaker properly programmed into GPS is capable of producing almost anything. But then, such an unknowable watchmaker is not much use in predicting outcomes, even if he is blind. Sounds rather like the creationists' problem.

In conclusion, it seems to me that there is indeed good reason to suppose that metaphysical assumptions have constrained vision in neo-Darwinian biology. Genomes that contain a high level of encoded morphological diversity in the form of error-checked coherent entities seem to appear with regularity. Neo-Darwinism can explain the exploration of such packages, but it has not proved that it can explain their origin. Based on uniform human experience, the simplest explanation for the appearance of a novel, dense pattern of information is an information-dense source. If available DNA templates seem inadequate, the alternative is a source of order exterior to the genome. Are there any known material sources of sufficient density to act as such sources other than human intelligence? Further, if no adequate material source suggests itself, is not the remaining logical explanation an immaterial source? Such hypotheses are excluded by the methodological assumptions of science. But—think the unthinkable—is that an adequate reason to reject the possibility? One cannot *logically* exclude a hypothesis of material inadequacy on the basis of one's *a priori* assumption of material adequacy.

Neo-Darwinism has been constructed (1) under a metaphysical

commitment to (global) materialism, (2) under the methodological commitment of science to use strictly material causal explanations, and (3) under the assumption that good science never lets a problem rest as "presently unsolved." It follows that in places where material explanations of cause are thin (problematic), they should be treated as anomalies waiting for a more complete (material) explanation rather than as mysteries, or as reasons for reviewing the adequacy of the methodological assumption. And certainly, due to the key role played by neo-Darwinism in the apologetic of metaphysical materialism, thin spots in that theory can be expected to be frequently overlooked even as scientific problems. When recognized, such anomalies are still likely to be shelved with the "best" material explanation attached, not declared unsolved. Indeed, according to Lightman and Gingerich (1992), such anomalies will probably not even be recognized as anomalies until a new paradigm able to explain them is proposed. If the assumption of global materialism is wrong, that might not happen until that assumption is rejected as necessary.

All of this may do as a working method for a materialist who has faith in God's absence. However, it does not justify telling the theist that, although God may exist (since science cannot prove otherwise), he is unemployed, since undirected material mechanisms have taken over his job. Assumed mechanisms are only assumptions, not proofs. (In any case, theists have never believed that *any* material event was undirected. How in the world could anyone demonstrate that *any* material event was not being directed?)

Science has proved neither that the material universe is undirected, nor that our material explanations are adequate. Therefore we should seriously re-examine the conclusions we have reached while working under the materialist agenda. *Has* anyone seen the emperor's new clothes recently?

REFERENCES

Amato, Ivan. "Capturing Chemical Evolution in a Jar." *Science* 255 (1992): 800.

Bakker, R. T. *The Dinosaur Heresies*. New York: William Morrow and Company, 1986.

Barton, N. H. and M. Turelli. "Evolutionary Quantitative Genetics: How Little Do We Know?" *Annual Review of Genetics* 23 (1989): 337-370.

Belyaev, D. K. "Destabilizing Selection as a Factor in Domestication." *The Journal of Heredity* 70 (1979): 301-308.

Boag, P. T. and P. R. Grant. "Intense Natural Selection in a Population of Darwin's Finches (*Geospizine*) in the Galápagos." *Science* 214 (1981): 82-84.

Borstnik, B., D. Pumpernik, and G. L. Hofacker. "Point Mutations as an Optimal Search Process in Biological Evolution." *Journal of Theoretical Biology* 125 (1987): 249-268.

Brooks, D. R., J. Collier, B. A. Maurer, J. D. H. Smith, and E. O. Wiley, "Entropy and Information in Evolving Biological Systems." *Biology and Philosophy* 4 (1989): 407-432.

Bryant, P. J. and P. Simpson. "Intrinsic and Extrinsic Control of Growth in Developing Organs." *Quarterly Review of Biology* 59 (1984): 387-415.

Burton, N. H. and G. M. Hewitt. "Adaption, Speciation and Hybrid Zones." *Nature* 341 (1989): 497-503.

Carroll, R. L. *Vertebrate Paleontology and Evolution*. New York: W. H. Freeman and Company, 1988.

Coates, M. I. and J. A. Clack. "Fish-like gills and breathing in the earliest known tetrapod." *Nature* 352.18 (1992): 234-236.

Dawkins, R. *The Blind Watchmaker*. New York: W. W. Norton and Co., 1986.

Desmond, A. *The Politics of Evolution*. Chicago: University of Chicago Press, 1989.

Endler, J. A. *Natural Selection in the Wild*. Princeton, NJ: Princeton University Press, 1986.

Erwin, D., J. W. Valentine, and J. J. Sepkoski, Jr. "A Comparative Study of Diversification Events: The Early Paleozoic versus the Mesozoic." *Evolution* 41.6 (1987): 1177-1186.

Gillespie, N. C. *Charles Darwin and the Problem of Creation.* Chicago: University of Chicago Press, 1979.

Gould, S. J. "Darwinism and the Expansion of Evolutionary Theory." *Science* 216 (1982): 380-386.

Hall, B. K. "Epigenetic Control in Development and Evolution." *Development and Evolution.* Eds. N. Holder, C. C. Wylie, and B. C. Goodwin. Cambridge: Cambridge University Press, 1983. 353-379.

Kingsolver, J. G. and M. A. Koehl. "Aerodynamics, Thermoregulation, and the Evolution of Insect Wings: Differential Scaling and Evolutionary Change." *Evolution* 39 (1985): 488-504.

Lande, R. "The Maintenance of Genetic Variability by Mutation in a Polygenetic Character with Linked Loci." *Genetic Research* 26 (1975): 221-235.

Larson, A. "The Relation between Speciation and Morphological Evolution." *Speciation and Its Consequences.* Eds. D. Otte and J. A. Endler. Sutherland, Mass.: Sinauer (1989): 579-598.

Lewin, R. "A Lopsided Look at Evolution." *Science* 241 (1988): 292-293.

Lightman, A. and O. Gingerich. "When Do Anomalies Begin?" *Science* 255 (1992): 690-695.

MacFadden, B. J. and R. C. Hulbert. "Explosive Speciation at the Base of the Adaptive Radiation of Miocene Grazing Horses." *Nature* 336 (1988): 466-468.

Marx, J. "Homeobox Genes Go Evolutionary." *Science* 255 (1992): 24-26.

Michaux, B. "Morphological Variation of Species Through Time." *Biological Journal of the Linnean Society* 38 (1989): 239-255.

Müller, G. B. and G. P. Wagner. "Novelty in Evolution: Restructuring the Concept." *Annual Review of Ecology and Systematics* 22 (1991): 229-256.

Pakula, A. A. and R. T. Saner. "Genetic Analysis of Protein Stability and Function." *Annual Review of Genetics* 23 (1989): 289-310.

Parsons, P. "Evolutionary Rates under Environmental Stress." *Evolutionary Biology* 21 (1987): 311-347.

Parsons, P. "Evolutionary Rates: Effects of Stress upon Recombination." *Biological Journal of the Linnean Society* 35 (1988): 49-68.

Parsons, P. "Evolutionary Rates: Stress and Species Boundaries." *Annual Review of Ecology and Systematics* 22 (1991): 1-18.

Paterson, H. E. H. "The Recognition Concept of Species." *Species and Speciation*. Ed. E. S. Vrba. Pretoria: Transvaal Museum Monograph (1985): 21-29.

Raff, R. A. and J. C. Kaufmann. *Embryos, Genes and Evolution: The Developmental-Genetic Basis of Evolutionary Change*. New York: Macmillan, 1983.

Rieppel, O. "Structuralism, Functionalism, and the Four Aristotelian Causes." *Journal of the History of Biology* 23.2 (1990): 291-320.

Rose, K. D. "Skeleton of *Diacodexis*, Oldest Known Artiodactyl." *Science* 216 (1982): 621-623.

Rose, K. D. "Climbing Adaptations in the Early Eocene Mammal *Chriacus* and the Origin of the Artiodactyla." *Science* 236 (1987): 314-316.

Roth, V. L. "The Biological Basis of Homology." *Ontogeny and Systematics*. Ed. C. J. Humphries. New York: Columbia University Press (1988): 1-26.

Sereno, P. C. and R. Chenggang. "Early Evolution of Avian Flight and Perching: New Evidence from the Lower Cretaceous of China." *Science* 255 (1992): 845-848.

Simpson, G. G. *The Meaning of Evolution*. New Haven, CT: Yale University Press, 1967.

Stanley, S. M. *Macroevolution: Pattern and Process*. San Francisco: W. H. Freeman and Co., 1979.

Stanley, S. M. "Rates of Evolution." *Paleobiology* 11.1 (1985): 13-26.

Templeton, A. R. "The Meaning of Species and Speciations: A Genetic Perspective." *Speciation and Its Consequences*. Eds. D. Otte and J. A. Endler. Sunderland, MA: Sinauer (1989): 3-27.

Thaxton, C. B., W. L. Bradley, and R. L. Olsen. *The Mystery of Life's Origin: Reassessing Current Theories*. New York: Philosophical Library, 1984.

Turelli, M. "Population Genetic Models for Polygenetic Variation and Evolution." *Proceedings of the Second International Conference on Quantitative Genetics*. Sunderland, MA: Sinauer (1988): 601-618.

Vrba, E. S. "Evolutionary Pattern and Process in the Sister-Groups Alcelaphini-Aepycerotini (Mammalia: *Bovidae*)." in *Living*

Fossils. Eds. N. Eldredge and S. M. Stanley. New York: Springer-Verlag (1984): 62-79.

Wake, D. B. "Homoplasy: The Result of Natural Selection, or the Evidence of Design Limitations?" *American Naturalist* 138 (1991): 543-567.

Wagner, G. P. "The Biological Homology Concept." *Annual Review of Ecology and Systematics* 20 (1989): 51-69.

Weber, K. E. "How Small Are the Smallest Selectable Domains of Form?" *Genetics* 130 (1992): 345-353.

Webster, G. and B. C. Goodwin. "The Origin of Species: A Structuralist Approach." *Constructional Biology.* Eds. H. Wheeler and J. Danielli. London: Academic Press, 1982.

Wellnhofer, P. "Archaeopteryx." *Scientific American* (1990): 70-77.

Wilkins, H. "Prinzipien der Manifestation Polygener Systeme." *Z. Zool. Syst. Evolutions Forschung* 18 (1980): 103-111.

Yoo, B. H. "Long-Term Selection for a Quantitative Character in Large Replicate Populations of *Drosophila melanogaster.1.* Response to Selection." *Genetic Research* 35 (1980): 1-17.

13a
Response to David L. Wilcox
Darwin Twisting in the Wind
Arthur M. Shapiro

SOME THINGS ARE SAID more precisely in one language than in any other. The French have a saying, *jeter de la poudre aux yeux*. The *Nouveau Petit Larousse* defines it as "causing someone to believe (something) by dazzling (them) with words or manners." It is a perfect description of Professor Wilcox's paper; that is, there's a lot less there than meets the eye. The biology in it is OK, but *so what*?

Wilcox begins by reiterating the focus of this symposium (albeit worded in a subtly different way, which could be significant but is not from my perspective, so I'll let it pass). He wants to know if the success of the "Darwinian paradigm" is due to its perceived metaphysical necessity or to its successes as a scientific explanation of empirical phenomena. He then makes an important distinction that suggests a sound logical structure to come: he decouples evidence for ancestry from mechanisms generating morphological novelty. (I suppose there are some people today who still confuse the two. Historically, this has been mostly a vitalist or orthogenetic error, and vitalism is clearly a manifestation of something other than ideological materialism; orthogenesis might or might not be strictly materialistic, depending on how the innate evolutionary "tendencies" are "explained." There is also an element of confusion in the claim by advocates of punctuated equilibrium that morphological novelty and speciation occur simultaneously. This is a somewhat different kind of confusion and at any rate is patently falsified by neontological data; it is the sort of error one would expect paleontologists to make, since the only "species" they can see are morphospecies.)

Very quickly, however, Wilcox lapses into debater's tactics by attacking a straw man of his own making.

"The first problem is a matter of simple logic," he asserts. "How can natural selection be a creator of new morphology when it does not write genetic messages, but only chooses among them?" Wilcox has the privilege of revealing philosophical secrets that have been well known since 1859. Whence comes the implicit equivalence of evolution (or the "Darwinian paradigm") with natural selection, such that natural selection is required to be a "creator of new morphology"? Again, "given a rabbit in the hat, a magician can pull it out—but how do rabbits get into the genetic hat? That, too, natural selection must answer if it is to be the genetic maestro." If natural selection were a financial commentator on the evening news, it might have to explain how the Federal Reserve regulates the money supply. But no sensible person expects it to do that. Who but anti-evolution debaters expects natural selection to be the "genetic maestro," anyway? By Michael Ruse's terminology, "ultra-Darwinians" do. I don't know any, but if they exist, they are probably busy writing economic treatises for the Cato Institute. Anyway, the omnipotence of natural selection is certainly not a logical necessity of evolutionary theory.

At several points in his essay Wilcox alludes to hierarchical phenomena. Hierarchy theory experienced a limited degree of trendiness in biological circles a few years ago, largely for sociological reasons. But that doesn't mean it isn't sometimes useful (though usually not). The levels-of-selection controversy has been a fecund one both conceptually and empirically. Although natural selection cannot be expected to explain everything, it actually has impacts on the genesis of variation that are subtle and once removed from the mutation process per se.

Several decades ago the Oxford ecological-genetics school correctly forecast that selection would tend to build up complexes of modifiers that would define allelic dominance by controlling expression in heterozygote phenotypes. They also developed the concept of the supergene, which involves tight linkage developed by selection of adaptive chromosomal reorganizations. This originally emerged from studies of the mimetic butterfly *Papilio dardanus*, dovetailed with studies of chromosome inversions in *Drosophila* by Dobzhansky and his collaborators, and later fed into various ideas about speciation, developmental genetics, linkage disequilibrium, and genomic constraints on design, some of which seem to bother Wilcox.

It was an easy step from there to the idea that DNA repair mechanisms were themselves subject to selection, which means that forward and back mutation rates can be seen as adaptive phenomena, rather than arbitrary "givens" in the system. (Because directed mutation is potentially so adaptive, its recent revival as a real possibility is no surprise; the idea is attractive not for ideological reasons but because if mechanisms exist for it to occur, a good Darwinian would predict that they would be selected for.) (When geographic races of *Drosophila* are hybridized, a short-term increase in mutation rates is sometimes seen, which is interpreted as a result of rendering heterozygous various loci involved in mutation repair in the different genomes.) All of this *is* evidence for the creativity of natural selection, but it does not add up to the claim of omnipotence that Wilcox thinks is required by the "Darwinian paradigm."

The early Darwinians were very open to a variety of mechanisms at work simultaneously in evolution. As is well enough known, evolution itself was much more popular than natural selection, and selection was in fact eclipsed for decades by a potpourri of what were seen as sexier explanations, including macromutation and various forms of vitalism. The triumph of selection was the fruit of the great success of theoretical population genetics in the hands of Fisher, Haldane, and Wright, and, paired with it, that of the empirical Ford-Kettlewell-Dowdeswell-Cain school at Oxford, whose theoretical mentor was Fisher.

Again, everybody knows that the overemphasis on micro-evolutionary process and the efficacy of natural selection inspired a reaction, triggered by Eldredge and Gould, Gould and Lewontin, in their critique of the "Panglossian paradigm" and the notion of punctuated equilibrium. It also included a revival of interest in Goldschmidt's premature synthesis, in the 1930s, of evolution and developmental biology under the rubric "physiological genetics." All of this was seized upon by critics of various stripes, from Fundamentalist Christians to Marxists, as proof that "Darwinism" was dead.

An engineering approach, stressing the properties of biomaterials in morphogenesis, developed; it was intended to explain innate constraints on morphology but helped briefly to fuel an essentially stillborn effort to transplant structuralist ideology from anthropology and linguistics into biology. Structuralism was correctly rejected by the vast majority of biologists because it was synchronic (and thus useless for addressing questions of ultimate causation, which are clearly diachronic) and because it summoned up memories of German idealistic morphology, which some of us would rather forget. We cannot, of course, because it so subtly and thoroughly interpenetrated so much of biology, and lies very near the heart of cladistics.

I rehash this recent history, lest anyone think there is anything new in Wilcox's philosophical complaints. There isn't. Even his pseudostructuralist language is derivative; and the recency of his citations merely shows how easily new wine can be poured into old bottles when evolution is at issue. Perversely, Wilcox almost seems to be saying that the more we know about biology, the more we need non-materialistic explanations. Why should that be true of the genome but not of the aurora borealis? Were the ancients better scientists because, knowing less, they speculated more than we do?

I will not respond at a technical level to Wilcox's citations because their content is so pervasively irrelevant to the matter at issue here. I will say that a proper cybernetic approach to genomic organization might be interesting (cf. Dembski), though its robustness would remain in question, given how little we actually know. That is, most of Wilcox's paper boils down to an assertion that this might be a problem worth pursuing.

That said, I want to talk twisters. Here in Texas tornadoes are a well-known, much-feared natural phenomenon. I happen to do meteorology as a hobby. Modern meteorology is highly technical and theoretical. It is just as mathematical as population genetics. Now here's a dirty little secret: we do not have a really satisfactory mechanistic understanding of how tornadoes work. We are, however, quite good at predicting where and when they are likely to occur. We can spot the conditions that spawn them and warn people who might be in their way, though we don't really understand *why* we can (Davies-Jones, 1992).

What does that say about our meteorological paradigm? Does it say that it "has been constructed *both* under a metaphysical commitment to (global) materialism, *and* under the methodological commitment of science to use strictly material causal explanation?" Does it say that "the seamless robe of meteorological (materialistic) stories would seem to have enough fundamental flaws to make it reasonable to question seriously the adequacy of the ruling metaphysical and even methodological paradigms? And if so, should we seriously re-examine the conclusions we have reached while working under the materialist agenda?"

You bet.

You won't catch me arguing that God can't make tornadoes anywhere and any time he pleases. If he chooses to stick 'em only onto certain kinds of thunderclouds under very predictable conditions, shoot, that's his right. After all, he's God.

So why do I have this nasty suspicion that if we got a tornado warning right now, Professor Wilcox would set aside his doubts about the "materialist agenda" in his rush for the cyclone cellar?

LITERATURE CITED

Davies-Jones, R. P. 1992. Tornado dynamics. In E. Kessler, ed., *Thunderstorm Morphology and Dynamics*. pp. 197-236. University of Oklahoma Press.

13b
Reply to Arthur M. Shapiro
Tamed Tornadoes
David L. Wilcox

ARTHUR SHAPIRO HAS more or less suggested that I have blinded the poor audience with a razzle-dazzle array of biological problems, a heterogenous display of difficulties too extensive to deal with individually in this forum. Further, his "tornado alley" illustration implies that my purpose is to inject the hand of God into science by introducing unknowns, "gaps" in scientific explanations. Neither objection is valid.

The diverse evidence I presented centers on a single, common problem: a complex and structured genome that is characterized by programmed and error-checked entities, a cybernetic base for biotic reality. Darwinian theory is based on a "bean-bag" view of genetics, on the additive effects of many small effect genes—or perhaps on occasional "macro" mutation. But, in the cybernetic model, those "beans" are best explained as adaptive buffers for the genetic goal-seeking machinery. Evidence for *adaptive* change does not naturally expand into evidence for *prescriptive* change as it accumulates.

Thus, it is not minor anomalies, the occasional genetic tornado, that neo-Darwinism cannot (yet) explain. Rather, it has failed to explain the fundamental realities of biological systems. It has failed to explain the core of the apple. Why then has it been considered an adequate (nay, a necessary and vital) explanation for all of biological reality?

I have no metaphysical necessity driving me to propose the miraculous action of the evident finger of God as a scientific hypothesis. In my world view, *all* natural forces and events are fully contingent on the free choice of the sovereign God. Thus, neither an adequate nor an inadequate "neo-Darwinism" (as mechanism) holds any terrors. But that is *not* what the data looks like. And I feel no metaphysical necessity to *exclude* the evident finger of God.

I conclude that the easy acceptance of neo-Darwinism as a complete and adequate explanation for all biological reality has indeed been based in the metaphysical needs of a dominant materialistic consensus. One *can* be a theistic "Darwinian," but no one can be an atheistic "Creationist."

Index

PUBLISHER'S NOTE:

A set of thirteen commercial grade videos covering the complete presentation of original papers and responses of all Symposium speakers is available. Also included are audience question and answer sessions and sessions of interaction between the Symposium participants.

Note: The interaction and question and answer sessions are not included in the content of this volume, but provide a rich exchange for further clarification and development of ideas.

To obtain information on the availability and cost, address inquiry to:

Foundation for Thought and Ethics
P. O. Box 830721
Richardson, TX 75083-0721